William Henry Thornthwait

A Popular Treatise on Photography

William Henry Thornthwait

A Popular Treatise on Photography

ISBN/EAN: 9783744675444

Printed in Europe, USA, Canada, Australia, Japan

Cover: Foto ©berggeist007 / pixelio.de

More available books at **www.hansebooks.com**

A POPULAR TREATISE

ON

PHOTOGRAPHY.

ALSO

A DESCRIPTION OF, AND REMARKS ON, THE

Stereoscope and Photographic Optics,

ETC. ETC.

BY

D. VAN MONCKHOVEN.

TRANSLATED BY W. H. THORNTHWAITE, PH.D., F.C.S.

Illustrated with   many Woodcuts.

LONDON:

VIRTUE BROTHERS & CO., 1, AMEN CORNER,

PATERNOSTER ROW.

1863.

PREFACE.

THE high repute of D. Van Monckhoven on the
Continent as a practical photographer, and the very
favourable reviews which his works have from
time to time received from the leading photographic
journals in this country, have led to the following
free translation of his "Traité Populaire de Photography sur Collodion." The alterations and emen-
dations which have been made in the original text,
it is hoped, will render the whole more immediately
available to the English reader, and acceptable as a
handbook of photographic art.

W. H. THORNTHWAITE.

April, 1863.

CONTENTS.

———◆———

A GLANCE

PRINCIPAL PHOTOGRAPHIC PROCESSES
NOW IN USE.

INTRODUCTORY.

AMONGST the various methods which have been from time to time proposed for the production of pictures by the chemical agency of light, and comprehended under the general term Photography, there are only four—the DAGUERREOTYPE, CALOTYPE, ALBUMEN, and COLLODION processes—which have been to any extent practically successful.

The photographic picture is obtained in the—

> *Daguerreotype* process, on polished metal plates;
> *Calotype*, on paper;
> *Albumen*, on a film of albumen on glass;
> *Collodion*, on a film of collodion on glass.

And it is an interesting fact that, although at first sight these processes do not appear to have any connection with each other, there nevertheless does exist one general principle of action, which will be obvious, even to the superficial observer.

THE DAGUERREOTYPE.—This process was discovered by Niepce and Daguerre. A silver, or silvered copper, plate, highly polished, is placed in the dark on a china dish, containing *iodine*. The vapour of this substance combines with the silver of the metallic plate in such a manner as to produce *iodide of silver*, a substance sensitive to light. In fact, it is sufficient to expose it behind a perforated card in the daylight for some considerable time to produce an impression; but should the time of exposure be much shortened there will not be any picture visible on the

layer of iodide of silver, although such really exists, and
can be *developed*, or made to appear. This can be effected
by placing the plate over mercury heated to about 148°
Fahr., so that it receives the vapours which arise there-
from, when an exact image of the perforated card will be
apparent in a very few seconds. There always remains a
certain quantity of iodide of silver which has not been
altered by light, because it has been protected from the
action of this agent by the opaque body which covered
the plate in certain places. If this iodide of silver were
not removed, it is easy to understand that it would become
changed as soon as exposed to the light; and therefore
it is necessary to dissolve it by a body which has received
the name of a *fixing* agent. Many substances dissolve
iodide of silver, such as the various *iodides, cyanides*, and
alkaline hyposulphites; but amongst these the most useful
is the *hyposulphite of soda.*

The Daguerreotype process just described will be perceived to essentially consist of a layer of *iodide of silver*,
the use of a *developer* to bring out the latent picture, and
a *fixing* agent for removing that portion of the iodide of
silver not acted upon by the light.

THE CALOTYPE process was the invention of Mr. Fox
Talbot, and, in consequence, is sometimes termed *Talbo-
type*. It consists in spreading, on a sheet of paper, first a
solution of iodide of potassium, and then a solution of nitrate
of silver. These two bodies, by their mutual reaction,
produce a yellowish-white, insoluble powder; then, by
washing the paper in water contained in a porcelain basin,
all the excess of nitrate of silver is carried away, and
finally there results a paper the pores of which are completely impregnated with iodide of silver. If this *iodised
paper* be exposed to the light, behind a perforated card,
the iodide of silver will become black at the places where
the light strikes upon it; or if a very short exposure
be given to the paper, on examination by yellow light no
image will be perceived. The picture can, however, be
developed by immersing the paper in an aqueous solution of gallic acid, mixed with a very small quantity of
another aqueous solution of nitrate of silver and acetic
acid. This mixture constitutes Mr. Talbot's *developer*.

At the end of some minutes the image shows itself, increases in vigour, and becomes very distinct ; the paper is then removed from the basin of gallo nitrate of silver, washed in water, and the iodide of silver not affected by the light is dissolved out with a solution of hyposulphite of soda.

The two processes of Daguerreotype and calotype, although employing such different substances as metal and paper, nevertheless have many points of resemblance, as in both the pictures are obtained on iodide of silver, requiring development and subsequent fixing; but when the respective pictures are examined, a very important difference will be perceived in their general aspect. A proof taken by the Daguerreotype will exhibit a counterpart of the original, with all the lights and shades correct. If it be a view taken with the aid of a *camera*, the sky will appear white, the trees a little darker, and the shadows black as in nature ; and, in the case of a portrait of a person standing before a white wall, the picture of the wall will be white, the hair and features differently tinted, and the dress black; consequently, it is usual to call this image *direct*, or *positive*.

It is quite different with the calotype picture; in fact, the view will have a black sky, the trees lighter, and the shadows white ; and, if it were the portrait, the wall black and the dress white; or, in other words, the picture will in every respect, as to depth of tint, be the reverse of the original, and for this reason is called *indirect* or *negative*.

Figs. 1 and 2 give an idea of these *negative* and

 *Fig.* 1.    *Fig.* 2.

positive images. Thus Fig. 1 represents a black cross on a white ground. If it be reproduced by the Daguerreotype an exact copy is obtained; whilst if the calotype process be used, the result will be an inverted image, or the cross will be white on a black ground; in fact a

negative image, as shown at Fig. 2.   Figs. 3 and 4 show
another example of these negative and positive proofs.*

There are many drawbacks against the employment of
the Daguerreotype process; for instance, the plate possesses a dazzling brightness, which forces the observer to
incline himself in some favourable angle to be able to
examine the details of the picture; besides, it must be
protected by glass, because the image which it bears on its
surface is destroyed by the least friction; lastly, and this

Fig. 3.

is above all its principal disadvantage, it only gives a single
image, whilst the other methods furnish an indefinite
number.   When a *negative* image is once obtained, it can
be employed to produce a series of other images, which
will also be reversed in relation to the negative, and consequently *positives*.   Take for example Figs. 3 and 4.
Suppose Fig. 3 the model to be reproduced, Fig. 4 will be
the negative image on the paper.   But if the latter be laid on
another *sensitive* sheet of paper, and exposed thus arranged

* See Note 1.

to the sun, its rays will pass through the white parts, and impress the sensitive paper which is underneath, but will be stopped by the black parts; and thus will the facsimile of Fig. 3 be produced. It will be understood that the same negative Fig. 4 can be used as often as required, and an unlimited number of positives analogous to Fig. 3 thus obtained.

From the necessity and manner of employing a paper negative to produce the required positive impressions, a very correct idea will be formed of the motives which have

Fig. 4.

led photographers to replace paper by a more homogeneous substance; however fine a surface paper may appear to possess, if it be examined by transmitted light it is always very uneven in texture, which circumstance greatly injures the delicate tints and sharpness of detail in the positive proofs.

THE ALBUMEN PROCESS.—The white of an egg, otherwise called albumen, is a transparent liquid, which, spread

on a plate of glass, leaves by evaporation a coating as
clear as the glass itself, so that when employed as a
photographic vehicle the most minute details are preserved
with perfect fidelity.

The manner of operating with albumen is exactly the
same as for the negative paper. In the albumen, properly
prepared, is dissolved a small quantity of iodide of potassium; this is spread on a well cleaned plate of glass; the
dried glass is immersed in nitrate of silver, exposed to the
light in a camera, developed as a proof on paper, and fixed
by hyposulphite of soda.

The advantage that albumen offers over paper, is the
production of finer details in the picture. It is not, however, at all an easy or sensitive process, and therefore
unsuitable for taking portraits, as at least ten minutes
exposure is required, even in a very good light, to produce an impression.

THE COLLODION PROCESS, which employs a film of
collodion spread on glass, possesses all the good qualities
of albumen, with the very great advantage of being, at
least, sixty times more sensitive, and withal easier of
execution. It is this process, and the subjects immediately connected therewith, that will be treated upon in
the following chapters.

CHAPTER I.

THE PLAN FOLLOWED IN THIS WORK.

IN this chapter it is intended to give a brief summary of
the subjects to be subsequently described in detail under
their several heads.

If a little pyroxyline or gun-cotton be immersed in
a mixture of about one part in volume of alcohol, and two
parts of ether, it will almost entirely dissolve. In order
to obtain a sufficiently transparent solution, it must stand
for twenty-four hours in a bottle well corked, and then
be poured out into another bottle, taking care not to dis-

turb the thick part which remains at the bottom. This clear liquid is *collodion*.

If a small quantity of collodion be poured on a glass well cleaned, the ether and alcohol will evaporate, and leave on the glass a transparent film. This film is very firm, and adheres strongly to the glass ; it is that which serves as a vehicle for the photographic materials; or, in clearer terms, it is that which is destined to form the surface which will retain the photographic image. It may easily be conceived that for the purpose of rendering this film sensitive to light, iodide of silver must be formed in its texture ; this is done by dissolving iodide of potassium, or some other iodide, in the collodion.

It is important to choose a proper iodide, but for the present purpose the use of a collodion containing iodide of potassium will suffice to trace what results and changes take place. On a well-cleaned glass plate pour a certain quantity of collodion in such a manner as to cause it to flow over its surface ; then incline the plate, that the excess of liquid may flow off. After the ether and alcohol are evaporated,—or, in other terms, after the plate has become dry,—a coating of pyroxyline will be obtained ; but this time it will have an iodide intimately mingled with it.

In proportion as more or less gun-cotton is dissolved in the mixture of ether and alcohol, a liquid of greater or less density is obtained, and consequently the thickness of coating of pyroxyline will also vary on the glass. The proportion of iodide added to the collodion also regulates the quantity which remains on the glass; from which it follows that it is not a matter of indifference what formula is employed in its preparation ; on the contrary, it is necessary to study with the greatest care the relative quantities of the chemical substances which constitute photographic collodion. These quantities will depend upon the temperature, and yet more upon the results desired to be obtained.

The glass plate having the collodion spread over it, is now *sensitised* by being dipped into a solution of nitrate of silver, which converts the iodide of potassium into iodide of silver, sensitive to light.

It need hardly be mentioned that the iodide of silver

being affected by light, the preceding operation of sensitising ought to be done in the *dark.* This word ought not
always to be taken literally; in photography, we understand by darkness a light too feeble to affect the sensitive
coating. A wax candle is generally used, or ordinary
daylight neutralised by a yellow glass; for a glass of
this colour prevents any action on photographic substances.

The sensitised plate now requires to be placed in a
camera, an apparatus composed of a box of wood and an
arrangement of lenses, which possess the property of
forming with perfect accuracy an image of any required
object on the sensitised plate. As a general rule, the collodion plate is left in this apparatus from ten to twenty
seconds, according to the brightness of the object; it is
then removed, and taken back into the dark room.

If at this moment the collodion film be examined with
attention, no trace of an image will be perceived, but it
can be made to appear in the same manner as has before
been mentioned, by developing with gallic acid. There
are also many other developers for bringing out the latent
image, such as pyro-gallic acid, proto-sulphate of iron,
proto-sulphate of uranium, &c. Whatever the developer
may be that is used, it is dissolved in water, and poured
over the coated surface of the glass. In a few seconds
the image appears as a negative, and the reduction is
allowed to proceed; or, in other terms, the proof left to
darken, until it is judged to be sufficiently distinct. The
glass is then plunged into water, which removes all soluble
substances, then into a fixing solution, such as cyanide of
potassium, or hyposulphite of soda, which dissolves the
semi-opaque coating of iodide of silver; finally, the glass
is washed in a current of cold water, and dried in the air.

If the proof be examined by transmitted light, it will
be found to be a true negative; that is to say, supposing
a view has been taken, the sky, the white houses, and
in general all objects strongly illuminated, are shown of
a black colour; while dark objects, on the contrary, appear transparent (Figs. 3 and 4).

The use of such a negative as before mentioned,
is to give a number of other proofs either upon glass
or upon paper; and if the tints be in good harmony

with the original model, a satisfactory picture will be obtained.

To understand the above requires a proper comprehension of the principle, that in spite of the opposition of lights and shades shown in a negative, with respect to a given model, there must nevertheless be preserved a perfect harmony between the tints. This phrase may appear obscure, but a well selected example will make it clear. Suppose a series of ten bands be fixed on a wall, of which the first is absolutely black, the last perfectly white, and the others of intermediate tints. The first will then be black, the second of a greyish black, the third a little less dark, the fourth still less, and thus lighter and lighter, to the perfectly white.

If a corresponding figure be reproduced on collodion, it will be found, if the negative be a good one, that these ten tints are completely in-
verted. In the place of the first black tint of the model, will be one perfectly trans-
parent on the negative, whilst the last will be of an opaque black, the intermediate ones having a regular gradation; for if it were otherwise, a proof taken from such a negative

Figs. 5 and 6.

would not represent the true shades of the original model. The Figs. 5 and 6 are intended to show this effect on three bands; but the result would be analogous for any larger series.

The conditions necessary to obtain this exactitude are, that the coating of collodion be of proper thickness, and sufficiently furnished with iodide of silver to yield a decomposition of such intensity as to produce a thick coating of reduced material; for if the light has not been able to decompose enough iodide in the coating of collodion, a perfect black can never be produced. *It follows, then, to obtain an intense negative, there must be employed a thick collodion strongly iodised, and a concentrated bath of nitrate of silver.*

The foregoing are the general details for producing

negatives upon collodion. This process, however, like that of the Daguerreotype, can be made to give direct or positive proofs at one operation; but in that case the picture requires to be viewed by reflected, instead of by transmitted, light.

All the operations necessary to obtain a negative upon collodion require to be performed in rapid succession. But if it be wished to delay the exposure and development of the picture for some time after sensitising the plate, it is re-coated with gelatine, or other suitable substance. The advantage of this method of operating, besides allowing some time to elapse between the exposure and development, is that it does not require the whole of the materials to be taken to the place where it is wished to operate; and consequently, although much less sensitive than wet collodion, is very applicable for views, &c. This method is known by the name of the Dry Collodion Process.

Some remarks on the stereoscope, an instrument particularly adapted for viewing photographic pictures, and full details for printing positive proofs from collodion negatives, and a few notes on optical photography and other photographic subjects, will complete the work.

The following is a list of the subjects treated on in the respective chapters :—

CHAPTER II.

ON THE PREPARATION OF SUBSTANCES REQUIRED IN THE MANUFACTURE OF PHOTOGRAPHIC COLLODION.

A MIXTURE of *alcohol, sulphuric ether,* and *gun-cotton* forms a liquid called *plain collodion,* to which is added, to render it suitable for photographic purposes, an iodide or bromide ; it is then termed *iodised* or *sensitised collodion.*

1. *Alcohol.*

Spirits of wine, or alcohol, is a liquid well known, and can generally be procured sufficiently pure for photographic purposes ; it boils at about 172° Fahr., and burns with a bluish flame without leaving a residue.   One hundred ounces by weight of alcohol measure about 125 fluid ounces, and 100 fluid ounces weigh about 80 ounces.   It should be perfectly clear, transparent, and absolutely free from any floating impurities; should it be otherwise, it must be carefully filtered.   The operation of filtering, applicable to other liquids as well as alcohol, is thus performed.   A circular sheet of *filtering* paper is first folded in two, as

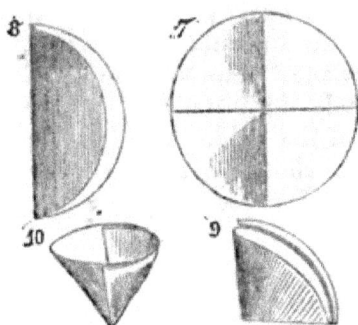

Figs. 7, 8, 9, 10. Method of folding Filtering
Papers.

Fig. 11. Apparatus for
Filtering.

represented by Figs. 7 and 8; then a new fold is made in the middle, Fig. 9; the filter is then opened out, as shown

at Fig. 10, and placed in a funnel, *b*, furnished with its
support, Fig. 11. The alcohol or other fluid
to be filtered, is poured carefully into the filtering paper, through which it will pass perfectly clear. Should the first portion that
runs through not be quite bright, it must be
returned to the filter. When a very rapid
filtration is desired, the filter paper may be
folded into a number of plaits, as shown at
Fig. 12, which affords a larger surface of paper for the
liquid to pass through.

Fig. 12. Plaited Filter.

The strength of alcohol is conveniently ascertained by
a specific gravity hydrometer. This instrument is formed
of glass, Fig. 14. It consists of a glass bulb with a glass
stem attached at the top, and a smaller bulb filled with
mercury, to serve as a counterpoise at the bottom. In
the stem is placed a graduated scale of specific gravities,
and the whole is so arranged that when
placed in pure distilled water, the instrument floats, and the surface of the water is
coincident with 0 or 1·000 on the scale. When
placed in alcohol or any fluid lighter than
water, the hydrometer sinks according to its
strength or specific gravity. A test-glass or
cylinder, Fig. 13, is used to hold a sample of
the alcohol or other liquid to be tested, and
care must be taken when the hydrometer is
placed in it that it floats perfectly free in the
fluid, and that no air bubbles attach themselves to its surface. The specific gravity is
then immediately indicated by noting the
degree cut by the surface of the fluid. Alcohol, suitable
for photographic purposes, should not have a greater
specific gravity than ·819. Its best strength is about
·803 to ·810.

Figs. 13, 14. Hydrometer and Glass.

2. *Ether.*

It is comparatively easy to procure alcohol of sufficient
purity for photographic use, to what it is to obtain pure
sulphuric ether. When manufactured on a large scale for
ordinary pharmaceutical purposes, there is generally so

little care taken, that the ether becomes contaminated with sulphovinic acid, aldehyde, or, worse than all, a peculiar principle resembling *ozone*, which is capable of decomposing iodides and liberating free iodine, therefore highly detrimental to its photographic action. The formation of this substance is much accelerated by the joint action of air and light; the knowledge of this circumstance is, therefore, of some importance to photographers, as it indicates a very necessary precaution to be taken to keep ether, and liquids containing it, particularly collodion, in well filled and closed bottles.

Sulphuric ether is a colourless liquid, and very volatile; when poured into water it floats about like oil, and a very small quantity is dissolved. It is much lighter than water, 100 ounces by weight of ether being equal in volume to 135 ounces by weight of water. Ether is tested as to its strength by the specific gravity hydrometer, in the same manner as described for alcohol on the opposite page. It should have a specific gravity of ·750 to ·720 to be of any use in photography.

In consequence of the highly volatile nature of ether, and its vapour being very explosive when mixed with atmospheric air, it is necessary, to prevent accidents, to avoid pouring ether from one vessel to another in a close room, or in proximity to a fire, or flame. As the vapour of ether is much heavier than the air, it naturally tends to fall, and therefore it is a proper precaution to take, when employing ether or fluids containing it—as collodion, for example—by artificial light, to have the source of light situated at some distance *above* the vessel from which the ether or collodion is poured.

Ether, if not sufficiently pure for photographic purposes, can generally be made available by the following method of rectification :—

Into a tall bottle, Fig. 15, is to be poured the ether to be purified, together with one-fourth of its volume of water, and the opening closed with a cork; the whole is then strongly agitated, and left to settle for some few minutes. Two layers of liquid will be perceived, the under layer being water slightly etherised, and the upper ether. The cork is now removed, and the shorter end of an ordinary glass syphon, having a small bore, and previously filled with water, is introduced through the neck of the bottle, and quite to the bottom of the liquid.

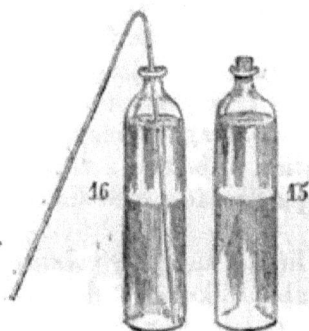

Fig. 16.   Fig. 15.

The smallness of the bore, and keeping the finger over the longer end of the tube, will enable the above to be done with facility, without the water from the syphon running out. The finger being removed, the syphon begins to act, and the etherised water from the bottom of the bottle is quickly drawn off. When the under layer has nearly disappeared, the orifice of the tube is again stopped with the finger, and the syphon removed.

A fresh quantity of water is now poured into the bottle containing the ether, which is again agitated and drawn off by the syphon as before explained.

This operation is called "washing," and the ether after this process is called "washed ether."

If bent glass tubes can be conveniently made or obtained, the following arrangement may be found more convenient than the ordinary syphon, it is shown at

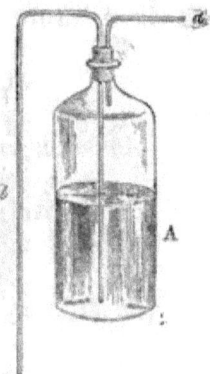

Fig. 17. Bottle with Syphon.

Fig. 17: A, the bottle where the ether and water is shaken together; it is furnished with a good cork pierced with two holes, in one of which is fitted a narrow tube (a) about $\frac{3}{8}$ths of an inch internal diameter, and in the other, a curved syphon tube (b), of which the shorter end inside the bottle reaches to the bottom. If the cork be properly fitted, it is only necessary to blow slightly through the tube (a) to cause the liquid to rise in the tube (b) and flow over. When nearly the whole of the underlayer of liquid has passed over, the syphon is stopped with the finger, the cork removed, and the fresh quantity of water added, and the operation gone through a second time.

The ether having been well washed, now requires to be dried and distilled; this is done by pouring the ether remaining in the washing bottle into a distilling vessel containing some few pieces of quicklime.

A convenient arrangement of apparatus for the distillation of small quantities of ether, is shown at Fig. 18; where larger quantities are operated on, the glass retort should be replaced by a vessel of zinc or tin plate. A is a small furnace for charcoal, B a vessel of copper or iron of some convenient form, to hold a small quantity of water, C a glass retort or other vessel, the opening of which is attached, by means of a cork, to a small leaden tube about the thickness of the little finger, and 1½ yards long;

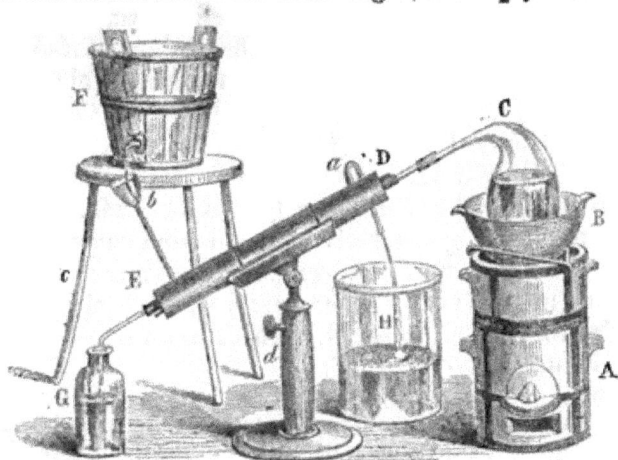

Fig. 18. Apparatus for distilling Ether.

a portion of this tube is surrounded by another about 2 inches in diameter, and ¾ths of a yard long; the top and bottom of this tube is closed perfectly water-tight round the smaller tube, it has also an overflow tube (a) at the top part, and a funnel and tube (b) at the bottom, through which a stream of cold water can be passed from any convenient vessel, as F, and discharged into the receptacle H. The end of the small leaden tube is bent so as to dip into a perfectly clean bottle (G); in every other respect the figure will convey a correct idea of the construction of the apparatus.

When about to be used, each separate part of the apparatus should be perfectly cleaned and washed out with water, and arranged as described and shown in the cut. The glass retort (C) or other vessel, is filled for about one-fourth its volume, with small pieces of quick-lime, and the washed ether poured on to it until two-thirds of the bulk of the retort is filled; the end of the leaden tube is then attached to the neck of the retort, and the refrigerator E D arranged in an inclined position, and firmly fixed by its support (d) so that the bent end of the tube dips into the mouth of the bottle, G, which is to receive the distilled ether. The whole being thus arranged, a small quantity of water is poured into the vessel B, so that the lower portion of the retort C is immersed in it, forming what is called a water bath; some lighted charcoal is now placed in the furnace A, and the water in the vessel B becoming heated, communicates its heat to the ether in the retort, which begins to evaporate, and in a short time drops of ether appear at the bottom of the leaden tube, and the distillation begins.* The water in the vessel B gets more and more heated as the bulk of ether in the retort diminishes, until no more drops are perceived to fall from the end of the tube; the fire is now removed, the apparatus separated, and the retort (C) or other vessel at once cleaned out, for should this be delayed, it becomes very difficult of performance.

The heat of the fire must be kept as much as possible from the bottle (G) containing the distilled ether, and a current of very cold water passed through the refrigerator E D, otherwise the ether vapour is not condensed. Sulphuric ether, rectified in the manner described, although not absolutely and chemically pure, is nevertheless well adapted for photographic purposes.

3. Gun-cotton.

Gun-cotton, also termed "pyroxyline," is nothing more than ordinary cotton combined with peroxide of nitrogen. It can be prepared by plunging cotton wool for a few minutes into concentrated nitric acid, then washed in water and dried; but in order to obtain a good pyroxyline for photographic purposes, a particular process must

be followed, and a rigorous attention paid to each
separate detail. Gun-cotton in appearance much resembles ordinary cotton, but it is heavier, and its fibres
break more easily ; it possesses also a slightly yellow tint,
which resembles that of raw cotton as imported into
Europe from the colonies. It is insoluble in water,
alcohol, pure ether, sulphuret of carbon, or chloroform,
but it dissolves in acetate of ethyle and methyle, methylic
alcohol, acetone, and also in alcoholised ether.

Pyroxyline burns with violence when brought in contact with any flame ; so much so as in many instances to
answer the purpose of common gunpowder.

The solution of gun-cotton in alcoholised ether is called
collodion, and is employed in surgery and photography ;
but for this latter purpose it requires to be specially
manufactured.

The following is the method of preparing gun-cotton for
photography, although we strongly recommend its being
purchased ready made, as photography being now so ex-
tensively employed, gun-cotton is prepared on a large
scale, and at a low price.

In a porcelain mortar is placed 2 ozs. of saltpetre in
fine powder, and over it is poured 3 ozs. by weight of
sulphuric acid of commerce. With the pestle, or a large
glass tube, the materials are well mixed, so as to obtain
a homogeneous paste. In this is immersed, in successive
portions, $\frac{1}{4}$ oz. of carded cotton, free from any mechanical
impurities. The cotton is pressed down with the pestle
until thoroughly wetted and imbedded in the liquid paste.
The mortar is then covered with a plate, to prevent the
nitrous vapours which are given off from vitiating the air
of the laboratory. It is also advisable to perform this
operation, if possible, in the open air.

The cotton is left in this mixture for ten minutes ;
the mortar is then placed in an inclined position, and
water poured into it, at the same time pressing the cotton
with the pestle so as rapidly to remove the excess of acid.
After washing for a half minute in this manner, it is taken
up with the hands and thrown into a wooden tub filled
with water, and well kneaded ; or else held under a water-
cock, and constantly worked about, and from time to time
pressed strongly between the hands. This washing should

be thoroughly done, until a portion of the cotton, when **put
in contact with blue litmus paper,*** does not produce a red
stain. It is then strongly pressed, and left to dry in the
air or in the sun, having previously spread it out thinly,
so as to present a large surface to the air. When the
cotton is dry it is preserved in glass bottles, well stopped.
Gun-cotton, thus prepared, very often gives traces of sulphate of potass; but as this substance is absolutely insoluble in ether and alcohol, it is of no importance.

Large quantities of gun-cotton should not be bought
or prepared at one time, as it appears to be liable to
decompose by keeping.

Gun-cotton, or pyroxyline, can be prepared according
to the formula given above from paper, linen, or hemp;
but these preparations have not been sufficiently studied
for us to recommend their employment in photography.

At the end of this volume (Note 3) some details are
given relative to the manufacture of gun-cotton on a large
scale by a mixture of nitric and sulphuric acids. In
general, the gun-cotton so prepared is less soluble than
that which has just been described; it, however, yields
an excellent collodion, especially adapted for coating large
plates, from its being very adherent.

Gun-cotton was discovered by M. Schonbein, a German chemist, in 1846. The photographic process which
employs collodion as its basis was first described by
Mr. Archer, in England, in 1851. M. Schonbein prepared
gun-cotton by steeping cotton in monohydrated nitric
acid. Afterwards M. Meynier discovered the advantage
of using a mixture of concentrated nitric and sulphuric
acids, and the method of preparation with saltpetre and

* Blue litmus paper is turned red by acids; red litmus paper is turned
blue by alkalies. These two papers can be bought ready prepared, or
they can be made in the following way:—Half a pound of litmus, in
small cakes, is boiled in an iron vessel with one quart of water for some
minutes, and then poured through a fine piece of linen, to separate the
undissolved portion. This solution is spread over paper by means of a
camel's hair brush, and the blue paper thus obtained hung over a cord to
dry. To make the red paper, a small quantity of vinegar is added to the
foregoing blue liquid until it becomes of a reddish colour. It is best to
cover the paper on both sides, and to cut it into small bands, which
should be kept in closed bottles, so as to prevent the action of acid or
alkaline vapours.

sulphuric acid is due to M. Marc Antoine Gaudin, calcu-
lator in the Bureau des Longitudes of France.

In connection with the method of preparing collodion,
presently to be described, will be indicated some other
important points as further guides to the selection or
manufacture of a good gun-cotton.

4. *The Iodides and Bromides employed in the Prepara-*
tion of Photographic Collodion.

A great number of iodides, bromides, and their com-
pounds have been at various times proposed for sensitising
collodion, but the formula most to be recommended is a
mixture of *iodide* and *bromide of cadmium.* In Note 4
will be found some remarks on the employment of the
iodides and bromides of potassium and ammonium.

Cadmium is now easily procured, almost in a pure state,
and at a comparatively cheap price. This metal is generally
found in commerce in small cylindrical ingots, about four
inches in length, and one-fourth in diameter. Its purity
can be known by its making a ringing crackling noise
when bent, like tin. If it bends with difficulty, and produces no sound when bent, it contains some other metals,
usually copper and zinc.

Iodine is a crystalline substance, having the aspect of
black-lead, or plumbago, volatile at a slight increase of
temperature, giving off purple vapours, highly corrosive,
and irritating to the eyes; it should always be preserved
in glass-stoppered bottles. It is obtained from the ashes
of burnt sea-weeds.

Bromine.—This substance is obtained from sea-water,
after all the common salt has been removed by boiling.
It is a very dense, dark red liquid; its vapour is highly
injurious and corrosive, and, from its great volatility, is
always kept under a stratum of water or sulphuric acid,
and in glass-stoppered bottles. Both iodine and bromine
are easily procured from any chemist.

Iodide of cadmium is thus prepared :—In a glass flask,
containing a quart of water, at first put in 8 ozs. of iodine,
and immediately after 4 ozs. of cadmium in small pieces.
The flask is placed on a stove, moderately heated, in such
a manner that the water in the flask shall be kept only

warm, not boiling. At the end of a few hours, especially if shaken from time to time, the liquid, from red, which it was at first, will become entirely colourless. Leave it to cool, and then filter. The cadmium that remains may be used for another operation.

The solution of iodide of cadmium thus obtained, is evaporated in a porcelain capsule. After a certain time crystals will appear in the liquid. It is then placed on a very hot stove, where all the water is driven off, and a dry mass obtained. The resulting substance is detached from the capsule with a knife, then reduced to a fine powder in a mortar, and finally preserved in a stoppered bottle.

The iodide of cadmium thus prepared is of a yellow tint, very soluble in water and alcohol, but less soluble in ether.

Bromide of cadmium is made by pouring 6 ozs. of bromine into 1 quart of water, contained in a stoppered flask; 4 ozs. of cadmium, in small pieces, are now added, and the flask closed. This mixture is left for some days, and very carefully shaken from time to time; the liquid gradually becomes discoloured, from the absorption of the bromine; when this takes place it is filtered and evaporated to dryness, as described for iodide of cadmium.

Bromide of cadmium is of a white colour, and less soluble in water and alcohol than the iodide. These substances, when prepared for sale on a large scale, are obtained beautifully crystallised, which may be taken as an evidence of their purity.

CHAPTER III.

PREPARATION OF PHOTOGRAPHIC COLLODION.

As the preparation of photographic collodion requires a considerable degree of nicety in the operations of measuring and weighing, it will not be out of place here to make a few remarks relative thereto that may be of some utility to the practical photographer.

Liquids are measured in glass vessels graduated into

ounces, drachms, and minims, the indicating signs and
relative quantities of which are shown
in the following table :—

1 pint contains 20 ounces, ℥ xx.
℥ j, or 1 ounce, contains 8 drachms, ℨ viij.
ℨ j, or 1 drachm, contains 60 minims.

Three of these graduated glasses
will be found necessary—one of the
shape Fig. 19, to hold 1 pint, and
divided into ounces; another of the
same shape, to contain 2 ozs. and divided into drachms,
and a small measure, of the form of Fig. 20, holding
2 drachms and graduated into minims.   When used for
measuring liquids, they should be held horizontally, on a
level with the eye, and the fluid poured in until its surface
reaches the line corresponding with the required figure
on the glass.   As these graduated glasses have lines corresponding with each other both at the front and back, the
proper position of the measure as regards its level is easily
shown.

Fig. 20.   *Fig.* 19.
Divided Measuring Glasses.

Fig. 21. Table Balance.

The table balance, Fig. 21, will be found the most convenient form for weighing quantities up
to 2 lbs.   It should have a set of weights
from ½ oz. to 2 lbs.

For weighing smaller quantities, the
hand scales, Fig. 22, is required.   The pans
should be made of glass, and there should
be a suitable set of weights, from ½ grain
to 2 drachms.   Glass is the best material
for the scale-pans; but if formed of metal,
it will be requisite, before proceeding to
weigh any chemical, to place a piece of
paper of equal dimensions in each pan, by
which arrangement the whole of the substance is con-

Fig. 22. Hand
Scales.

veniently removed after weighing, and any injurious action
avoided.   When fluids are required to be weighed, a
glass or other vessel to hold the liquid, is first accurately
balanced or counterpoised, and then the weighing done
in the ordinary manner.

The preparation of *Iodised Collodion* requires the
following substances :—

> Ether, sp. gr. ·720 .... 3 ounces.
> Alcohol, sp. gr. ·805 .... 1½ ,,
> Gun-cotton.. 16 grains.
> Iodide of Cadmium in powder .... 18 ,,
> Bromide of Cadmium   ,, ... 6 ,,

The gun-cotton is first put into a suitable glass bottle,
afterwards the iodide and bromide of cadmium and the
alcohol.   The mixture is then strongly agitated to dissolve the salts of cadmium, and to open the fibres of
the cotton, and facilitate its subsequent solution.   The
ether is now added, and the whole again well shaken
until the cotton is dissolved, when the bottle is closed
with a good cork, and left to settle for twenty-four hours,
after which the clear and limpid portion is decanted into
small glass bottles for use.

From the circumstance that collodion containing the
iodiser as above is liable to decompose, it will generally
be found preferable to prepare the collodion and iodising
solution separately in the following manner.

Plain thick Collodion, without iodides or bromides, is
first prepared as follows :—

Into a bottle of about one quart capacity is placed—

No. 1.

> Gun-cotton............................ 450 grains.
> Alcohol .... 7 ounces.
> Ether...................................... 25 ,,

These materials are very strongly agitated together,
and left to settle for some days.

A solution of iodide and bromide is also prepared
thus :—

No. 2.

> Alcohol .... 3½ ounces.
> Iodide of Cadmium .... 154 grains.
> Bromide of Cadmium ...... 54 grains.

The iodide and bromide of cadmium should be ground very fine in a porcelain or glass mortar with a small quantity of the alcohol, then the remainder added, and when the salts are dissolved, the whole carefully filtered. The clear solution must be preserved in a well-stopped bottle.

To prepare *Sensitised Collodion*, pour into a four-ounce bottle—

Thick Collodion (No. 1.) .... 1 ounce.
Solution of Iodide and Bromide (No. 2.) ... 3 drachms.
Alcohol .... 6 drachms.
Ether .... 1¾ ounce.

Immediately after the bottle has been well shaken, so as properly to mix the ingredients, the sensitised or iodised collodion is ready for use, but is improved by being kept a few hours after sensitising before being applied.

Both the plain collodion and the alcoholic solution of iodide and bromide of cadmium can be preserved for any length of time without deterioration, provided they are kept separate and in well-closed bottles; and the sensitised collodion can, therefore, be prepared when required with great facility.

The formula for sensitised collodion just given will be found to work best at a mean temperature of 60° Fahr. When the weather is very hot a little more alcohol must be added, and the quantity of ether slightly diminished; and, on the contrary, when the atmosphere is very cold the alcohol may be diminished, and the ether increased.

When a glass plate is coated with collodion, a larger quantity is poured over its surface than is really required, the surplus being received in a bottle; it will therefore be easily understood that after a considerable number of plates have been coated, the proper relative proportions of the constituents of the collodion will have been disturbed by evaporation, and that both ether and alcohol must be added to bring it to its normal condition. It must be borne in mind in making the required addition of ether and alcohol, that ether being much more volatile than alcohol, a larger relative proportion will have evaporated, and consequently a larger quantity of ether must be used than of alcohol.

If the layer of collodion on the glass appears too thin, a little of the thick collodion, No. 1, must be added ; if, on the contrary, it be too thick, and in consequence does not spread evenly over the glass, it can be diluted with a small quantity of a mixture of two parts of ether and one of alcohol.

If the collodion film detaches itself from the glass plate after being sensitised in the silver bath, it indicates that a larger proportion of ether is required, or that the gun-cotton employed is not suitable, in which case recourse may be had to a gun-cotton prepared according to the formula given in Note 3, which produces a very adhesive collodion, especially applicable when glasses of large dimensions are employed.

If too much alcohol be added to the collodion, the film is liable to become detached, and the coating itself has a wavy uneven appearance. If there be too much ether the layer is very adhesive, but it is difficult to get the collodion to spread itself evenly over the plate, especially if it be of any large size.

Collodion ought to be preserved in well-stopped bottles, but it is indifferent whether bottles with glass stoppers or furnished with good corks be employed ; they ought, however, as much as possible, to be kept quite full, and in the dark.

Iodised collodion is never good the first day of its preparation ; it must be kept for at least two days to acquire all its properties ; neither must it be kept too long, for impressions are then taken with it less rapidly.

Collodion is very unstable ; sometimes without any apparent reason it becomes slow in producing impressions ; at other times it changes to a red colour at the end of a few days. In this latter case recourse ought to be had to a new preparation, and care taken to ascertain if the materials previously employed were sufficiently pure. The colour of good collodion is commonly of a very light lemon colour, although sometimes completely colourless.

A very clear and perfectly settled collodion must always be made use of. The following is an excellent little apparatus for pouring out collodion free from sediment.

A very tall and narrow bottle must be procured of the form indicated by Fig. 23. The cork *a* is pierced with two holes made with a round file; two small glass tubes are fitted to it, of which A goes a little way through the cork, and the other is bent down in the form of ∩, only one branch is shorter than the other. The longer branch is dipped into the collodion at a short distance from the bottom. After a certain quantity of collodion has been used for several hours, what remains should be poured into the bottle by lifting the cork *a*, which must be again replaced. The next morning the collodion will be perfectly settled, and by blowing into

Fig. 23. Glass apparatus for decanting Collodion.

the tube A, the clear liquid passes by the tube D *a* B, from the extremity of which it is received into a proper bottle.

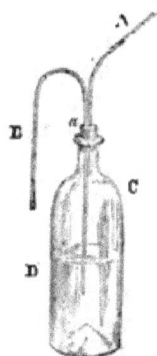

Care must be taken that the surface of the collodion be lower than the extremity of the tube B; should it be otherwise the tube must be raised by causing it to pass through the cork, or else the collodion would continue to flow after the required quantity had been decanted.

CHAPTER IV.

CLEANING GLASS PLATES, ETC.

GLASS plates for photographic purposes are employed with ground edges—that is to say, the glasses, after being cut with a diamond, are ground on their edges by means of a fine file, aided by oil of turpentine. The object of this operation is to prevent the operator from cutting himself upon the sharp edges of the glass.

In place of crystal sheet, ordinary patent plate, or even flatted crown glasses may be used, especially for

the smaller-sized (under half-plate). Patent plate being
more perfectly polished than flatted crown, is also more
easily cleaned, and therefore preferable.

Whatever be the kind of glass employed, the following
is the method which should be pursued in cleaning the
plates whether they have been used before or not :—

A mixture of equal quantities of nitric acid and water
is made, and the glasses covered therewith on both sides
by the aid of some cotton-wool fastened to the end of a
stick ; and as each plate is successively treated in this
manner, they are placed against the wall to drain.

Instead of nitric acid a solution of carbonate of potash,
of the strength of one pound to one quart of water, may
be advantageously employed. The liquid acts energetically upon plates that have been previously used, and
is free from the objection to which nitric acid is open,
namely, that of staining the hands yellow.

It is always a good plan to clean a number of plates at
the same time—as many as twenty, for instance—because it
is necessary to allow the alkaline or acid solution to act for
at least an hour. The plates are then submitted to the
action of a strong current of water, and rubbed at the same
time on both sides with a sponge, in order to remove all
impurities. It is almost unnecessary to add that they are
finished by allowing the water to flow in all directions,
copiously, on the inclined plate, to carry off every trace of
dust. The glasses are then placed in a grooved frame, similar

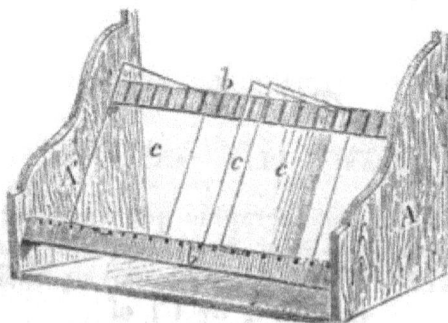

Fig. 24. Support for Cleaned Plates.

to Fig. 24, which precludes the necessity of description,
where they are allowed to drain and dry. The photo-

grapher should furnish himself with several of these draining frames of different sizes, for they will be found very convenient.

The glass having been cleaned, is not yet sufficiently pure to receive the collodion: it is further necessary to resort to a more complete polishing. For this purpose an oak plate-holder, of the form A B, Fig. 25, and of suitable size should be used. At the end B is placed a piece of wood about the thickness of a crown-piece. A

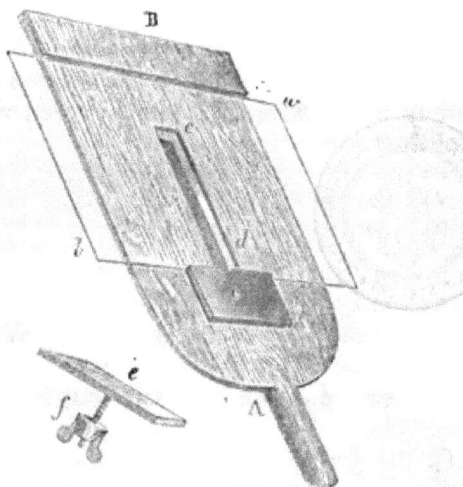

Fig. 27.   *Fig.* 26.   *Fig.* 25.

Tripoli Bottle and Plate-holder.

groove, c d, admits of a second piece, e, fixed underneath by a clamp-screw, Fig. 27, to fix the plate a b, of any size within the dimensions of the plate-holder.

In addition to this, there should be near at hand a box containing *powdered tripoli*, and a bottle furnished with a tube of small bore, Fig. 27, containing alcohol. The glass being fixed on the plate-holder, Fig. 27, is well dusted with tripoli, a few drops of alcohol are poured upon it, and then, by means of a little cotton-wool, or *papier Joseph*, the mixture is rubbed all over, carefully avoiding contact between the fingers and the glass. The rubbing is repeated, but without pressing too hard, until the alcohol has entirely evaporated, and the plate is dry. The excess of adherent tripoli being removed by a

dry linen cloth set apart for this especial purpose, the
final polish is given with an exceedingly dry buckskin or
chamois leather, which should also be used exclusively for
this operation.

The plate is then removed from the plate-holder, its
edges and back wiped, and placed in a grooved plate-
box. Care should be taken that that side of the glass
which has been polished for the reception of the collodion
film, should in each case face the same way, in order to
avoid errors in pouring on the collodion.

The plates should never be cleaned more than twelve
hours before using, especially if the box in which they
are contained is likely to be carried about, as in that case
they would again become covered with minute particles
of dust.

CHAPTER V.

ON THE DARK ROOM, AND PHOTOGRAPHIC LABORATORY.

In the practice of photography, three different rooms are
necessary : one well lighted, in which the sitter is placed ;
another of commodious dimensions, which serves the
purpose of a laboratory ; and a third made quite dark, in
which are performed all those operations which may not
see the light of day.

The laboratory should be of such dimensions as are
adapted to meet the wants of the photographer. Amateurs
generally take plenty of room, because in most cases it
costs them nothing ; but professional photographers are
oftener obliged to content themselves with a small apartment, which frequently serves the double purpose of dark
room and laboratory.

However this may be, it is best to have a large room,
where preparations can be made, positive proofs printed,
&c., on the ground floor, well furnished with drawers
and tables.

The dark chamber ought to be, on the contrary, very

simple.   Two or three tables are sufficient, and the light
should either be entirely excluded by pasting black paper
over the windows, and the operations conducted by the
light of a candle or a gas jet, surrounded by a square
lantern of yellow glass, or else, as often preferred, the dark
room is so arranged that the light comes exclusively
through a frame of yellow glass about 10 inches by 8 inches,
and this covered with a sheet of very thin white paper, in
order to impede the passage of the direct solar rays.   A
hinged frame is fitted in front of this square of yellow
glass in such a way as to admit of its being totally or
partially covered, in order to diminish or increase the
amount of illumination at pleasure.

Fig. 28. Dark Chamber.

It is necessary to select glass of a deep yellow colour,
and avoid the paler kinds.

The drawing (Fig. 28) represents the dark chamber
which has been used for many years by M. Monkhoven.

The entrance is at the bottom of the room, and the
door, for greater security, covered with a black drapery.
A cistern of water, furnished with a tap, supplies the
necessary means for washing, and underneath this is

placed the sink, with a pipe to convey the waste water outside. On one side should be placed the dishes used for nitrate of silver bath, and for sensitising albumenised paper, and on the other those which appertain to fixing, and other operations incompatible with the nitrate of silver solution. Shelves should also be fitted up to support funnels while filtering, bottles, &c.

A second reference to Fig. 28 will show the arrangement by which the amount of light is regulated; it will be seen, in fact, that by lowering the yellow frame, which is shown in that position in the drawing, the room becomes inundated with light. This frame need be closed only when the plate is immersed in the nitrate of silver bath, and during the development of the image. When the required intensity has been attained in the last operation, the plate is washed and the frame lowered in order to see more clearly. If the sun shines on this window, it is absolutely necessary to paste over it some thin paper to stop the too direct rays.

CHAPTER VI.

COATING WITH COLLODION AND SENSITISING THE PLATE.

THE nitrate of silver bath is prepared by dissolving—

Pure **Nitrate of Silver** .... 2 ounces.

IN

Distilled or Rain Water .... 24 ounces.

This liquid is filtered and poured into a gutta-percha tray. The collodion, spread upon the glass, is plunged into the nitrate of silver bath; the film whitens by the transformation of the iodide of cadmium into iodide of silver, which is sensitive to light, and it is in this state that it should be exposed in the camera.

Such is a general statement of the process to be now described somewhat in detail.

The collodion ought to be contained in a bottle with a wide mouth (Fig. 29), which mouth should always be carefully cleaned before pouring the collodion on the glass.

The box containing the cleaned glasses being placed in the dark chamber, a plate is withdrawn, and the dust removed from the polished side by means of a large badger-hair brush; then holding the plate by one corner (Fig. 30), in the right hand, the collodion is poured upon it from the left hand, commencing at the corner B. It is then allowed to flow by inclining the plate from left to right, and finally draining back into a separate bottle from the corner D, Fig. 31. If, at this moment, the plate be examined in a particular light, it will be seen that it is covered with an infinity of small ridges in the

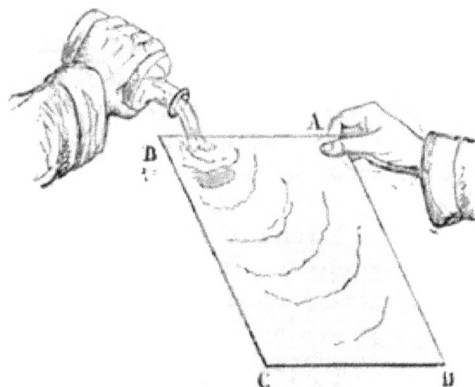

Fig. 30. Coating the Plate.

direction B D (Fig. 30); but on inclining the plate rapidly from left to right, these ridges will disappear. As soon as this happens, the plate is ready to be immersed in the nitrate of silver bath; but it is always advisable to wait a few seconds (and how many, experience alone can indicate precisely), so that the film may be sufficiently "set." There are certain characteristics by which it may be determined if the right moment has arrived for sensitising the plate; these are :—

1. If the plate is immersed in the bath before the

collodion has been allowed time to "set," the film will become detached in fragments; and, in this case, it will be necessary to filter the bath.

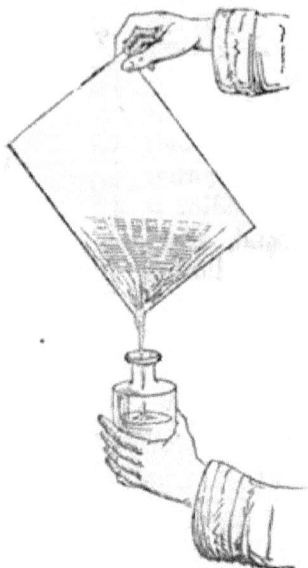

Fig. 31. Receiving the Collodion into a separate Bottle.

2. If the immersion take place at the right time, the film whitens gradually.

3. If, on the other hand, the film be allowed to become too dry before immersion, it whitens instantly, and a good proof is never obtainable under these circumstances.

Up to a certain point it is not difficult to determine when the plate should be immersed in the nitrate of silver bath. It will be found if the plate is looked at in a certain light, that it has assumed a dull, un-polished appearance. This, which is the right moment for immersion, will be arrived at in summer probably in about twenty seconds, while in winter it will probably take sixty seconds.

Fig. 32. Apparatus for filtering Nitrate of Silver Bath.

Nitrate of silver bath is very easy to prepare. It is quite as well for those whose operations are conducted on a somewhat extensive scale to have a considerable quantity of bath solution, and keep it in a large bottle to which is adapted a funnel and filter, Fig. 32. The same filter will last a long time; and, on leaving work, the nitrate of silver bath which has been used during the day is poured into the filter, and in this way a solution in proper condition is always maintained at hand.

For the nitrate of silver bath, dishes in gutta-percha, porcelain, or wood with glass bottoms, are used, Figs. 33, 34, 37. Gutta-percha dishes are, perhaps, most frequently employed,* but as vertical baths in glass and gutta-percha are also used, we shall say a few words in reference to them.

The reason why we recommend a gutta-percha dish in preference to one in porcelain is, that the film itself can be better seen, as also the impurities which float in the solution.

There are several methods of immersing the plates in the nitrate of silver bath; and as this is one of the most important points in the present chapter, we will pause to consider it somewhat in detail.

Figs. 33 and 34. Dishes in Porcelain and Wood with Glass Bottom.

The dish containing the nitrate of silver bath being much larger than the plate, is raised at one end, Fig. 38, in order that the solution shall accumulate at the opposite end; the plate, previously coated with collo-

Fig. 37. Gutta-percha Dish.

dion, is placed against the other edge, and held there by means of the finger or a silver wire hook; lowering then the plate by a continuous motion, and allowing the dish at . the same time to assume the horizontal position, the liquid flows at once, and without stopping, over the whole of the plate: after this the dish should be raised and lowered during a minute or so. Then introducing underneath the plate a hook made of flattened silver wire, Fig. 35, and by its means withdrawing the plate from the bath, it will be seen to be covered by a number of veins which show that the nitrate of silver bath

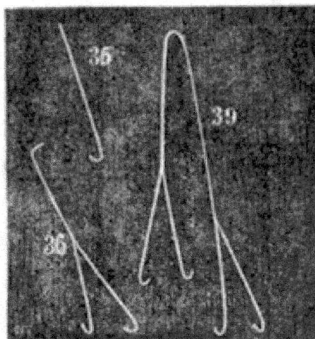

Figs. 35, 36, 39. Dippers.

* This remark is correct only as far as regards the practice of continental photographers,—in England glass and porcelain baths are always employed.

has not thoroughly penetrated the film, and that, therefore, the plate should be raised and lowered alternately
until the silver solution flows evenly and smoothly over
the whole surface of the film.

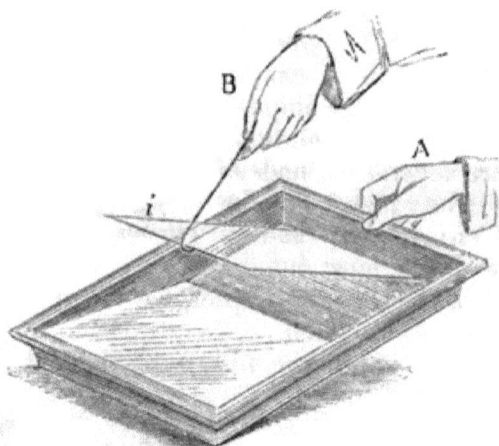

Fig. 38.

It is at this point that the plate should be withdrawn,
the fingers being covered with India-rubber finger-stalls,
in the absence of which
box-wood forceps, or an
American clip, should
be used, as the nitrate
of silver blackens the
hands very strongly
(Note 5). The plate
is allowed to drain from
the excess of adherent
nitrate of silver solution,
and is then placed in
the camera frame.

This method of ope-
rating requires only a
small quantity of nitrate
of silver bath solution,
which is accordingly
exhausted in a propor-

Figs. 40, 41. Vertical Glass Bath.

tionately shorter time. The plan we are about to describe
requires, on the contrary, a large bulk of solution, which

has the advantage of becoming very slowly exhausted. Moreover, a vessel may be used which is only just large enough to allow the plate to be immersed and withdrawn with freedom.

The solution is contained in a vertical bath, either of glass or gutta-percha, Fig. 40; and by means of a hook or dipper of gutta-percha, which is introduced on the lower side of the vessel, the plate is immersed by one continuous motion into the fluid, raising and lowering it alternately as before. The plate is finally withdrawn ready for exposure, when the ridges or lines thereon have disappeared. In Note 6 are given details for the preparation of nitrate of silver bath.

Another plan we can much recommend is as follows:—Two silver hooks are first procured, made of flat sheet silver, and bifurcate, of the shape shown in Fig. 37. The nitrate of silver bath being contained in a dish (Fig. 42) of gutta-percha, the plate is held between the two hooks, the film being uppermost, and plunged at one stroke beneath the liquid, taking care at the same time that one end is immersed before the other, for unless this be done the liquid will spirt out of the dish. The plate being once fairly covered by the solution, one of the hooks is removed; and with the other the plate is raised and lowered, in order to

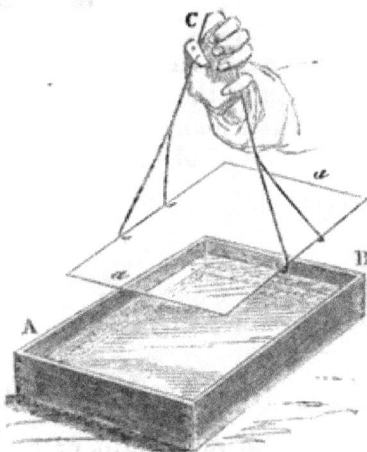

Fig. 42.

get rid of the veins or greasiness of which we have spoken above.

Instead of two hooks, it may be found more advantageous to use the two combined in one, as shown in Fig. 42. We may observe, in passing, that the silver hooks may be replaced by others made of whalebone. In order to make these, it is necessary to hold them in the flame of a spirit-lamp until they bend, and to allow them to cool, maintaining pressure in the proper curve by means of the fingers until quite cold.

Whichever method is followed in immersing the plate in the nitrate of silver bath, it is indispensably necessary that it be the result of *one steady and continuous motion;* for if this be not the case, lines will be formed upon the film, which will become apparent very soon in irreparable stains.

To preserve the hands from the action of nitrate of silver, some amateurs use India-rubber finger-stalls, or gloves of the same material; but such apparatus will be found very inconvenient, and those who intend to be successful in photography should make up their minds before going into it, to sacrifice the delicacy of their hands and the whiteness of their shirt-cuffs.

We presume it is hardly necessary to observe that, at the time of sensitising the plate, the room should be made quite dark—an operation easily accomplished by moving the shutter in front of the yellow glass until the light is almost entirely excluded. With a little practice this will be found comparatively easy.

As soon as the plate is placed in the camera-back, the window and door may be opened, and the bath covered with a plate of glass, in order to preserve it from dust.

CHAPTER VII.

PHOTOGRAPHIC CAMERAS, LENSES, ETC.

THE common camera-obscura is supposed to have been discovered by Baptiste Porta, about the year 1590. It may be simply described as a box, at one end of which is fixed a piece of ground glass, and at the other a convex or magnifying-glass, mounted in a sliding tube to regulate the *focus.*

All convex or concave glasses are called lenses; the focus of a convex glass or lens is the distance between the glass when exposed to the sun, and the point or spot of light where the rays unite (Fig. 43). (See Note 6.)

If the distance between the magnifying lens and the ground glass of an ordinary camera be regulated to the focus as described, and the lens directed towards some distant objects, it will be seen, on so shading the ground glass with a black cloth

Fig. 43.

placed over the head as to prevent any lateral rays of light falling on it, that the image of those distant objects are clearly represented *reversed* on the ground glass. If the camera be now directed to some near objects they will not appear distinct, and the lens will require to be drawn out further from the ground glass before they are shown with perfect sharpness. This adjustment of the lens is called *focussing* the image.

It will be found impossible, however, to focus the whole of the objects perfectly; there will always be certain parts of the picture which want distinctness. But if a piece of cardboard, having a small hole in it, be placed a short distance in front of the lens, the image on the shaded ground glass will become much more distinct and sharp, or, in other words, the picture can be better focussed. Any similar arrangement to that described is called a *stop* or *diaphragm*.

The foregoing remarks will explain the principles and construction of the *photographic camera*, which essentially

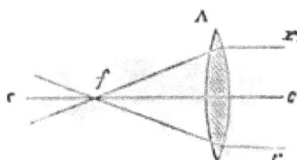

Fig. 44.   Ordinary form of Photographic Camera.

consists of two parts—the *lens* or *objective*, and the *box* or *camera*. However, it will be easily understood that these two apparatuses, to be rendered suitable for the various purposes of photography, must of necessity require more special and complicated construction than has been described.

The ordinary form of a photographic camera, Fig. 44, consists of a box, B, in which slides another box, A, holding the frame C with the ground glass. For the purpose of fixing the sliding part A, a board, D, is fastened to the box B in which is a groove, as shown in the drawing. On the lower side of the box A is fixed a plate with a thumb-screw passing into the groove. Thus, within certain limits, the two boxes A and B can be adjusted to various lengths, and fixed at the required focus by the aid of the thumb-screw. The lens is placed in front of B.

This form of camera has generally a single sliding-box, A; but sometimes, however, for the sake of portability, they are constructed with several slides. The material of which these cameras are made is usually mahogany or walnut wood.

The *tripod*, or *camera stand*, for supporting the camera, is usually of two kinds—one for travelling, the other for use in the operating room for portraits, &c.

The travelling stand, Fig. 45, consists of a strong wooden or metal triangle, upon which are made to slip three feet, which for greater strength and firmness are made double. These feet, for convenience in

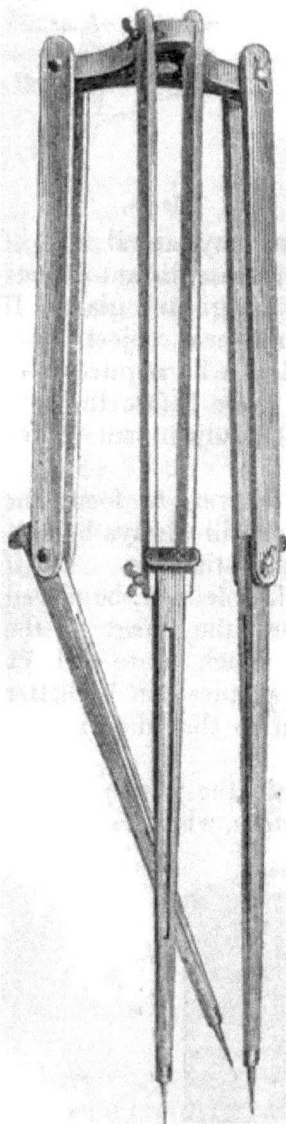

Fig. 45. Tripod Stand.

travelling, can easily be doubled up, and removed from the triangle by unscrewing the joints.

There are many different models of tripod stands for travelling, but they are all more or less on the principle of the one just described, and are made lighter or stronger as may be required for the size of camera. A black

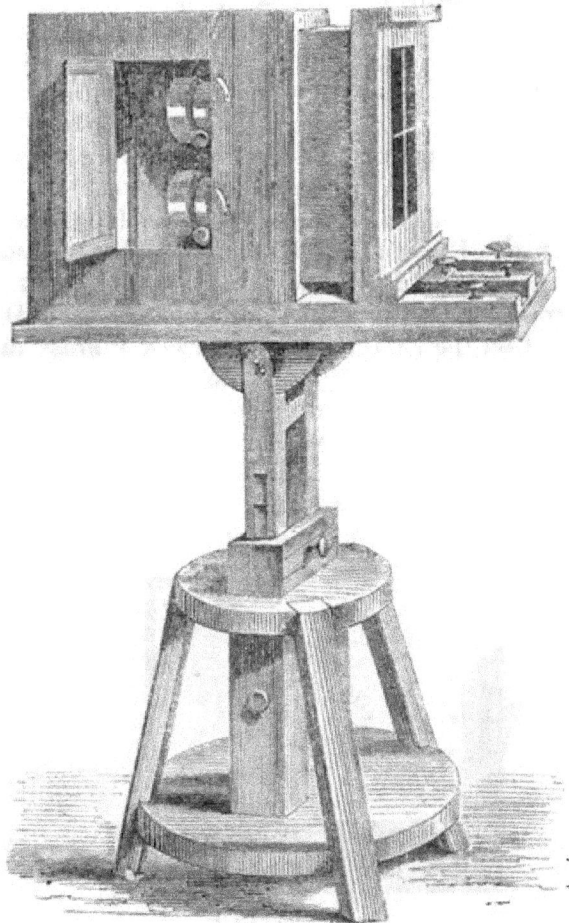

Fig. 46. Stand for Operating Room, and Camera for " Cartes de Visite."

cloth, so constructed as to cover over the extended tripod stand, forms a convenient dark chamber for changing sensitised glasses from or to the camera back when dry collodion plates are employed.

The stand used in the operating room or studio, not

requiring to be moved any distance from place to place, may be of a much stouter and heavier construction than the ordinary tripod stand. The form of camera stand represented Fig. 46, combines solidity with facility of adjustment. The upper board supports the camera, and in order to allow of a tilting motion is connected with two half circles of wood, which can be fixed in position by a thumb-screw to an upright sliding frame, moving in a socket of the firm tripod stand. This sliding frame serves to adjust the height of the camera. Sometimes this form of stand has a winch, with rack or chain, to facilitate the raising and lowering of the camera. This addition is very convenient, especially when a heavy camera is employed.

A convenient form of camera for travelling, and suitable for plates 12 inches by 10 inches, and larger sizes, is represented at Fig. 47. B is a fixed box, pierced in front with a round hole, to which is attached the lens. Underneath, and on the side (c) are two pieces of brass, in each of which can be fixed a thumb-screw, as seen at a. The sliding body A has also two similar pieces, one at d, the other underneath, each having a thumb-screw. The top of the tripod (which is seen in the figure) has a hollow, in which a long piece of wood b can be fixed; this has a groove in the middle, to admit the two thumb-screws.

Fig. 47. Camera without a Tail-piece.

By this arrangement, on turning the screw a, a very great degree of firmness is given to the part B of the camera; also to the tail-board b, and the top of the tripod. The sliding body A can be drawn out or in, and fixed as the operator wishes. The instrument can also be used on its side (the two pieces of brass which

are seen at c and d serve for this purpose), enabling upright views to be taken with the same facility as long ones.

This form of camera is more convenient than the ordinary form, Fig. 44, especially for taking large views ;
besides which, the immoveable tail-board of the model,
Fig. 44, prevents its being conveniently used on the side
for upright pictures.

Another form of camera, suitable both for the operating
room and for travelling, is represented, Fig. 48.   It will

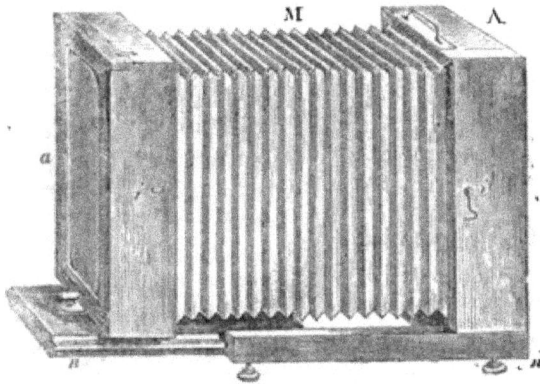

Fig. 48. Bellows-body Camera.

be found extremely light and portable.   The construction of this form of camera will be easily understood from
the cut.   M is a square bellows body, connected with the
wooden frames A, a.   The frame A, holding the lens, is
firmly fixed to a board n, n, which can be lengthened or
shortened at will.   The frame a holds the ground-glass
frame and dark slides.

Whatever may be the form of camera employed, the
general principles to be observed in its use are the
same.   For the purpose of illustration, suppose a camera
mounted on its stand, as shown in Fig. 55, p. 49, and
directed towards a person who stands for a model.   On
drawing out the sliding body more or less, a point is
reached where the proper focus is obtained.   As before
mentioned, the head of the operator and the top of the
camera must be covered with a black cloth, so that the
sharpness of the image on the ground glass can be properly
examined. Fig. 55 represents an operator focussing a person
who is placed before him. If the ground glass be replaced

by a glass plate, covered with sensitised collodion, an image will be obtained representing the model. The proper placing of the sensitised plate is accomplished by means of a frame represented in Fig. 49, and which is always sold with and forms part of the camera. This frame, or camera back, is represented open, in order to make the figure better understood. The glass plate, coated with collodion, after having been taken out of the silver bath and well drained, is placed in the camera frame, with the layer of collodion towards the shutter *a* (which has been previously closed); then the door *b* is shut and fastened with hooks, so that the sensitised layer is thus preserved from the

Fig. 9. Camera, or Dark Frame.

light. The ground-glass frame is then removed from the camera, and the frame (Fig. 49) put in its place. If the thin sliding board or shutter *a* be now raised, the sensitised surface of the glass plate, being exactly in the same position as occupied by the ground glass, receives the same image. When it is thought that the light has acted long enough, the shutter *a* is closed, the frame removed, and taken into the dark room, where, on opening the door *b*, the glass plate is removed, and submitted to an operation to be described under the title of *developing the image*.

The inner portion of the frame that holds the sensitised plate has projecting corners of silver wire, so arranged that the plate only touches at these parts. Thus the woodwork of the frame is in a great measure protected from the corrosive action of the nitrate of silver; and if the precaution be taken of placing a piece of blotting-paper on the lower corners and back of

Fig. 50. Glass Frame.

the plate, previously to closing the door *b*, the risk of drops of nitrate of silver injuring the frame, and falling during its transport from the dark chamber to the operating room, will be avoided.

With a camera of a certain size it is very often required to take smaller pictures, as well as the full size the lens will produce. This is accomplished by having some extra frames of wood (Fig. 50) fitting into the larger frame, and carefully adjusted, that when the plate is placed in one of these extra frames its surface shall be exactly coincident with the position of the ground glass upon which the focus was obtained; if otherwise, it would be impossible to produce a sharp and distinct picture.

The glasses employed in photography are usually of patent plate, although good ordinary glass may be employed for very small sizes. The glass plates are kept in grooved wooden boxes (Fig. 51).

Fig. 51. Plate Box.

There are two kinds of photographic lenses—those called *single lenses*, for views; and *double*, or *combination lenses*, for portraits.

The following are the diameters and focal lengths of portrait and view lenses suitable for the various sized glasses usually employed :—

FOR PORTRAITS.

	Size of Pictures. Inches.	Diameter of Lenses. Inches.	Focal Length. Inches.
Ninth size	$2\frac{1}{2}$ by 2	$1\frac{1}{4}$	$3\frac{1}{4}$
Quarter plate . . .	$4\frac{1}{4}$,, $3\frac{1}{4}$	$1\frac{3}{4}$	$4\frac{1}{2}$
Ditto for rapid action .	$4\frac{1}{4}$,, $3\frac{3}{4}$	$2\frac{3}{4}$	5
Third size ditto . .	5 ,, 4	$3\frac{1}{4}$	7
Half plate	$6\frac{1}{2}$,, $4\frac{3}{4}$	$2\frac{3}{4}$	7
Whole plate . . .	$8\frac{1}{2}$,, $6\frac{1}{2}$	$3\frac{1}{4}$	$9\frac{1}{2}$

FOR PORTRAITS AND GROUPS.

Size of Pictures. Inches.	Diameter of Lenses. Inches.	Focal Length. Inches.
9 by 7	$3\frac{1}{4}$ & $4\frac{1}{4}$	11
10 ,, 8	$4\frac{1}{4}$	13
12 ,, 10	$4\frac{3}{4}$	15
15 ,, 12	$5\frac{1}{4}$	19

FOR VIEWS.

Size of Pictures. Inches.			Diameter of Lenses. Inches.		Focal Length. Inches.
6	by	5	$1\frac{3}{4}$		8
7	,,	6	2		10
9	,,	7	$2\frac{3}{8}$		14
12	,,	10	3		16
16	,,	12	4		24

The *single*, or *view lens*, is composed of an achromatic lens, mounted in a tube of brass (Fig. 52), as shown in the figure. D, the brass cover of the lens. F, a small brass tube, holding diaphragms of various sizes. C, the brass tube, having the lens at A screwed within it, and also furnished with an outside screw, which allows the whole mounted lens to be fixed to the ring of metal A, which is attached to the front of the camera.

Fig. 52. Single, or View Lens.

The view lens is used in the following manner :—It is fixed to the camera by being screwed into the brass ring ; the cover D is removed, when the image may be focussed. All the diaphragms should be removed, except the one to be employed. The smaller the opening of the diaphragm, the sharper the image ; but then a longer time will be required to produce an impression. Thus, a diaphragm of $\frac{1}{4}$ inch requires four times as long exposure as one of $\frac{1}{2}$ inch ; but then the picture produced is much sharper. It is thus that the operator can, when required, sacrifice sharpness of detail for rapidity of action.

The lens requires from time to time to be cleansed with a soft wash-leather ; but in remounting the same in its cell, *it is of the greatest importance that the convex side of the lens should be towards the sensitive plate.**

The *compound*, or *portrait lens*, is composed of a greater number of lenses. It has a compound lens in front, which is very thick, being formed of two cemented together, like the view lens, and two other lenses at the

* View lenses are sometimes mounted, with the diaphragms or stops fixed in the large tube holding the lens, which either slides in another tube or is actuated by rackwork, to facilitate the obtaining a correct focus.

back. Whenever the lenses are removed to be cleaned, or for any other purpose, it is important that they be replaced in exactly the same order as when purchased, otherwise the picture produced can never be good and sharp. Fig. 53 shows the form of the double, or com-

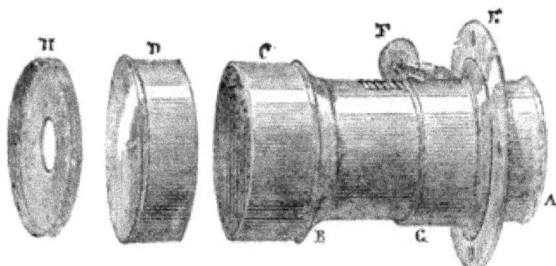

Fig. 53. Double, or Compound Lens.

pound lens. D is the cover, or cap; B G, a double tube, sliding one over the other, capable of being adjusted by turning the milled head F connected with the rackwork. Its use is, after the focus has been roughly obtained, by drawing out the sliding body of the camera, to finally adjust the same with the greatest ease and nicety. C is a portion of a larger tube, adapted to screw upon the tube B, and serves to protect the lens from lateral rays of light; B, A, the ends where the lenses are fixed. G, the tube, having a screw at the end for the purpose of attachment to the ring E. This ring of metal is fixed by screws upon the front of the camera. H, a diaphragm, or stop, fitting the interior of the tube C.

The above are the chief points of difference between the two varieties of lenses known as the *single* and *double*. The brass mountings of both forms are blacked inside, to prevent internal reflection, and should this dull black coating be at any time rendered imperfect, it can be repaired by the application of a mixture of lamp-black and gum-water.*

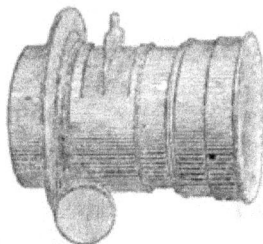

Fig. 54.

* In the most recently improved form of compound lens, the sliding body is pierced so as to allow a series of diaphragms, or stops, to be introduced between the two lenses, as shown in the illustration, Fig. 54. This arrangement is more correct in principle, and produces better results, than the simple external stop.

With respect to the optical differences of these two forms, and their special uses, it is to be noted that the single lens, although of the same size as the double combination, does not give so brilliant an image upon the ground glass, because the diaphragms intercept an enormous quantity of light. The double lens, on the contrary, has two lenses, combined in such a way that a diaphragm is not necessary; consequently, the images produced are well illuminated. By stopping off a single lens, the sharpness of the image is rendered perfect, whereas with the double combination such a degree of sharpness cannot be obtained; therefore the first is employed for views and buildings, and the latter for portraits.

Another consideration also influences the selection. It is easy to be understood that a sensitive coating is impressed more easily in proportion to the brilliancy of the light striking upon it. Thus, a lens of 3 inches in diameter, with $\frac{1}{2}$ inch stop, will require one hundred times longer exposure than a double combination of lenses of the same diameter to produce the same picture. Now, is it not easy to expose for as long a time as may be required for a view, landscape, flowers, or inanimate objects, so as to obtain, what ought to be the chief aim, extremely fine detail in the picture? In most cases, therefore, although rapidity can be gained by enlarging the stop, results will show it had better be avoided. With the single lens just mentioned, a view, illuminated with the sun, can be easily obtained in twenty to thirty seconds, and if there be any persons in the view, they will without doubt be copied. If, with the same single lens, an endeavour be made to take a portrait of a person placed in the shade, it will require from three to five minutes. Is it possible to remain so long without moving? Certainly not. Now the double lens will allow such a portrait to be taken in ten seconds; and although it gives a picture not quite so sharp as the single lens, it nevertheless is greatly preferable, and would in reality produce a sharper picture, for with the single lens no one could remain perfectly immoveable during a sufficient time.

In certain cases it is also necessary to use a stop with the double combination lens; but instead of a stop of $\frac{1}{4}$ to $\frac{1}{2}$ inch for a lens of 3 inches diameter, one of from 1 to $1\frac{1}{2}$ inch is all that is required. Fig. 54 shows such a stop at H ; it is placed in the hood C of the lens, and is chiefly employed when a group of portraits is required to be taken. For the purpose of more clearly comprehending the reason for the use of a diaphragm when taking groups, &c., with a double or compound lens, it may be remarked that, *the more extended the space occupied by an object, the more light will there be on the ground glass;* but at the same time, *with the double lens there will be less sharpness.* An intermediate course is therefore adopted ; that is to say, a diaphragm is employed, whereby a loss is incurred in the matter of exposure for the sake of gaining sharpness and definition.

Generally speaking, a single lens gives so perfect a definition to the whole of the object, that the operation of obtaining a good focus is extremely easy. It is not so, however, with the double combination lens; therefore, as a rule, the head of the sitter should be carefully focussed, and a little of the details sacrificed, if necessary, to obtain this point.

Compound lenses are manufactured and sold for taking pictures of a certain size ; this must be understood to mean the utmost of their capabilities, and those photographers who may be desirous of obtaining first-class portraits should always employ a larger lens than is absolutely required. The following is a list of compound lenses particularly adapted to produce the finest portraits of the given dimensions :—

Size of Picture. Inches.	Diameter of Lens. Inches.	Length of Focus. Inches.
$4\frac{1}{4}$ by $3\frac{1}{4}$	... $2\frac{3}{4}$ ...	7
$6\frac{1}{2}$ „ $4\frac{1}{4}$	... $3\frac{1}{4}$ ...	$9\frac{1}{2}$
$8\frac{1}{2}$ „ $6\frac{1}{2}$	... $4\frac{1}{4}$ ...	11
10 „ 8	... $4\frac{3}{4}$ ...	15

A new form of lens has lately been constructed for views, consisting of two lenses, which allows a great degree of sharpness to be obtained with more flatness of field than the ordinary single lens. They are termed *orthoscopic* or *caloscopic*. In a special note (Note 8), will be given a description of some of the principles of optics, a proper understanding of which will render more clear many of the foregoing details.

For taking "carte de visite pictures" the best size of compound lens that can be employed is $2\frac{3}{8}$ inches in diameter and 5-inch focus.

CHAPTER VIII.

ON THE GLASS ROOM IN WHICH THE SITTER IS PLACED,
AND THE RULES TO BE OBSERVED IN TAKING PORTRAITS
AND LANDSCAPES.

To take an *artistic* portrait, or to choose the most favourable point of view for a landscape, requires an artistic
taste not to be acquired by reading, but allied in its character to a natural instinct, of which instruction only
develops the germ, while practice simply modifies and
perfects the details. The remarks on the subject matter
of this chapter will therefore be of an essentially practical
nature.

The general detail and arrangement of an operating
room suitable for photographic purposes, is represented
in the cut on succeeding page. It should be in some
elevated position, either on the roof of the house or on a
platform specially erected for the purpose. The side next
the *south* should be entirely closed, whilst the other,
towards the *north*, is glazed. The sheets of glass employed should be of moderate thickness, as a protection
from storms of hail, &c., and as white as possible—rather of
a bluish tint than approaching at all to a green or yellow
colour. It is of importance to attend to the *colour of
the glass used for the operating room*, for, should it be of
a green or yellow tint, a considerable amount of actinic
rays* are cut off, and the exposure necessary for a good
portrait is greatly augmented.

The glass rooms should be furnished with curtains
having cords attached to them, by means of which the
too energetic action of the light may be moderated, and
proper *direction* given to it. In the figure it will be
seen that three sets of squares are shown—the upper one
(of ground glass), which is parallel with the floor or
ceiling; the lower one, which replaces the wall; and the
middle one, which is on the slope.

---

* Those rays of light which produce chemical action, and are found
at the blue and violet end of the spectrum.

If now, with the object of producing an artistic effect, it be desired to inundate the front of the model or sitter with light, the lower and middle set of curtains should be more or less closed. If a lateral lighting be desired, the curtains of the upper window should, on the contrary, be

Fig. 55. The Glass, or Operating Room.

closed. The good taste of the operator must, however, guide him as to the best disposition of the light.

On the right of the engraving, Fig. 55, is represented the shaft of a column, against which the model may lean; also a balustrade, with a landscape painted in distemper;

D

and near the middle a white marble chimney-piece of the style of Louis XV.; all of which accessories, or others of a similar character, will be found to aid in imparting a general appearance of elegance to the resulting picture.

The colour of the walls exercises a marked influence on the result. They should not be painted either red, yellow, or green, for these colours have a very weak photographic action, and throwing around them, as they do, tints of their own colour, tend to prolong very materially the time of exposure. Violet and blue colours are preferable, but as they produce whites in the print, and as a wall painted either deep blue or violet produces a result exactly similar to a white wall, they should not be used for the background, or that portion behind the sitter. Bluish grey is a mixed tint which, on the whole, yields the best results, and is a colour with which the whole of the glass room may be painted, except that, according to taste, some parts may be more or less deep than others.

It will be found very convenient to have several move-able backgrounds, each painted with a different depth of colour, so as to be used according to the colour of the dress, &c., of the sitter, and thus produce the most effec-tive contrast.

Oil colour is very disagreeable on account of its reflec-tion. It will be found best to employ a mixture of slaked lime, litmus, and lamp-black with which the whole of the room may be painted, simply varying the proportion of black. The same colour will serve to paint the floor of the room; but if a carpet be employed instead of paint, it is equally important that it should be of a greyish tint.

In reference to the best colour for dresses to be worn by sitters, the same remarks apply as have been made re-specting that of the glass room; that is to say—neutral tints, analogous to grey, violet, and blue, come out well, while red, yellow, and green, yield results of an oppo-site character. By increasing the exposure, however, in some cases, and diminishing it in others, the undue predominance of any particular tint can be materially diminished.

When one or more persons are to be taken, they should be allowed, in the first place, to assume an easy, natural position, and then, by placing behind each a *head-rest*, in such a way as to retain them in the position chosen, the required steadiness of the upper portion of the body is secured.

It is not necessary that the sitter should press the head too strongly against the head-rest, but on the contrary, it should only lightly touch; because too great a pressure restrains the respiration, thus imparting to the sitter an appearance of constraint and uneasiness.

There are two kinds of head-rests. One in iron, or of iron and wood, is represented in Fig. 55. The lower part consists of a tripod and tube of brass weighing about 40 lbs., which serves to prevent vibration in the upper portion, which is applied to the head. This consists of an iron tube having one piece forked, and capable of being adjusted and fixed by a screw in any desired position.

These head-rests are made to stand on the ground; but Fig. 56 shows one constructed so as to be attached to a chair.

This form of head-rest is generally made of hard wood :—*e* is a flat piece of wood, to be adapted to the back of a chair; two clamping screws, *f f*, are attached to it ; *i i* a grooved board which comes behind the chair. It may be easily raised and depressed, and is fixed in any desired position by turning the screws *f f*. *b b*, a double-jointed piece for adapting to the position of the head, and capable of being fixed by means of the screws shown in the figure. The whole of this piece may be raised and lowered, and fixed at *d*. *a* is a moveable forked piece, against which the sitter leans. It will be easily perceived that the back of the chair must come between the pieces *e* and *i i*.

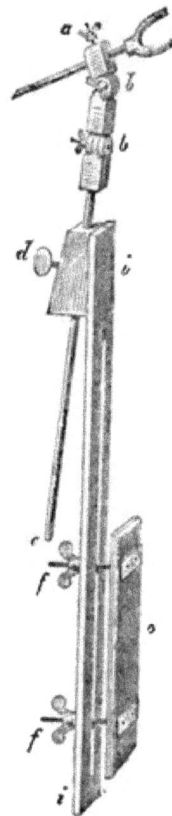

Fig. 56. Head-rest.

Figure 55 shows also the manner of obtaining a focus, in connection with which may be observed that too great an inclination should never be given to the camera ; as a rule it is best to have the lens about the height of the chest. A very slight inclination of the camera will then be sufficient to get a correct image of the sitter from head to foot.

In taking a portrait from a sitting position, it is best to lower the camera-stand a little, and thus avoid too great distortion.

The nearer the camera is brought to the sitter the longer the exposure ; and inversely, the further the camera is removed, the shorter the exposure. And it is in this way that the time of exposure may be varied from one second to three hundred. But as a general rule, in taking a full-length figure in summer, the plate should be exposed twenty seconds, while a sitting position will require thirty seconds. In winter these times of exposure should be increased one-half.

The following is a general summary of the operations involved in taking a portrait. The direction and amount of light are the first things which claim attention, then the attitude of the sitter. Focussing is the next operation, during which the sitter should be requested to keep still, though not maintaining perfect immobility. The plate should now be prepared in the dark room, on returning from which any alteration which may have taken place in the pose of the sitter during the absence of the operator is corrected, and efforts made, by cheerful conversation, to induce an agreeable expression. Then, when everything seems in order and ready, a final and rapid adjustment of the focus is made; the focussing-glass is withdrawn and replaced by the camera-back, containing the sensitive plate. The lens is now uncovered and the plate exposed, the necessity of complete stillness having been previously enjoined on the sitter, explaining, however, at the same time, that he may breathe in the ordinary way, and if necessary *wink*, but *not move* his eyes from the spot where first directed. The time of exposure having expired, the lens is covered and the slide of the camera-back closed, and the development proceeded with in the dark room.

With reference to the means to be employed to estimate the time of exposure in seconds, it will be found best to read the time from a good watch with a second-hand, though the same object may be attained by suspending a leaden or wooden ball by means of a cord 39¼ inches long. If the pendulum so constructed be made to oscillate, it will be found to mark seconds of time with sufficient accuracy.

It has been already mentioned in chapter seven, that lenses of a different construction are necessary for landscapes from those which are used for portraits. For views and architectural subjects a single achromatic lens is sufficient, but for portraits a double combination is necessary. It has also been stated that the form of camera-stand, Fig. 46, and cameras, Figs. 47 and 48, are those most applicable for landscape photography.

The rules which can be given for taking views are much more simple than those for portraits. In point of fact, success depends mainly on the taste of the operator in selecting the landscape which he desires to reproduce, and the particular point of view from which it is seen. The focussing is accomplished in the ordinary manner.

It is only necessary to add, that views are taken by the wet as well as the dry collodion process. With the latter all that is required is a grooved plate-box, containing some prepared plates, a camera-stand, and a large black cloth with which to cover the legs of the camera-stand, when it becomes necessary to replace an exposed plate by one which has not received the luminous impression. With the wet collodion process, a photographic tent, or a light tent carriage, is indispensable; and the operator must also take with him all the paraphernalia of dishes, baths, bottles, &c., which form the necessary furniture of a dark room.

Notwithstanding these difficulties, the wet collodion process is to be preferred for the reproduction of architectural subjects and landscapes near great cities. But for a long voyage, the dry collodion process is certainly the best.

In the wet collodion process, the nature of the result (good or bad) is known at once on the ground; while with dry plates the character of the picture is not ascertained until the development is effected, which almost invariably takes place at a great distance from the locality in which the view was taken.

In large cities and flat countries it is very easy to have
a tent or light carriage of waterproof cloth, carried, or
drawn if need be, by a porter or guide—an arrangement
which offers the advantage of allowing the operator to
stop wherever he pleases, and conduct the work with
great facility.

Generally, before taking views, a preliminary visit is
made, with the object of ascertaining the best points of
view, and on such occasions the iconometer, or view
meter, Fig. 57, is found
very convenient. It re-
quires to be expressly
constructed for each
focus of lens and size
of camera, and resem-
bles very much in ap-
pearance an opera-glass.
It consists of a small
lens, a camera, of the shape of an opera-glass, and a square
focussing glass. By turning the lens towards the view it
is represented reversed on the ground glass; and in this
way the operator can judge whether his large camera
will take in the whole or what portion of the required
view.

Fig. 57. Iconometer.

Instead of the iconometer with lens and ground glass,
a more simple one may be employed, which in many
respects, however, is similar to that indicated in Fig. 57.
The observation is made through the front opening, and
on the large circle behind (to the left in the figure) a
rectangle is described equal to that which is yielded by
the lens attached to the camera. It is then only necessary to observe what objects are included in the field of
vision, in order to ascertain what will be reproduced on
the ground glass. The iconometer with lenses is, however, the most convenient for general use, as it affords an
opportunity of knowing whether in the case of taking an
architectural subject, for example, the operator is or is
not too close, for then the vertical lines incline towards
a point, like the furrows in a horizontal field.

CHAPTER IX.

DEVELOPMENT OF THE LATENT IMAGE OBTAINED IN THE CAMERA-OBSCURA.

As soon as the proper time has elapsed for exposure, the sensitised plate must be withdrawn from the light by closing the shutter of the camera-back, which is then taken into the dark room. After carefully closing the shutter in front of the yellow-glass window, the plate is removed from the frame. If it be now examined, there will very rarely be any traces of a picture ; it can, however, be made apparent by covering the surface which has been exposed to the light with a solution of some substance capable of reducing the salts of silver to the metallic state.

Among the reducing agents which can be employed in photography, may be mentioned protonitrate of iron, sulphate of iron, pyrogallic acid, protosulphate of uranium, protosalts of osmium and titanium, &c.; but of all these, pyrogallic acid will be found the best for developing the negatives, and protosulphate of iron for obtaining direct positives on glass.

Pyrogallic acid is a white crystalline solid, without smell.* It is soluble in water, alcohol, and ether, and its solution in either of these menstrua becomes rapidly decomposed on exposure to the air and light, especially if the solution be alkaline. This substance alters very soon even when it is dry. If a small amount of moisture be present, however, the decomposition proceeds with increased rapidity, being expedited by the foreign matter which it contains. Thus it is better—first, to keep the pyrogallic acid bought in shops in small stoppered bottles ; secondly, to keep it in the dark ; thirdly, never to prepare with it more developing solution than is likely to suffice for a day's consumption.

It is usual to add to the pyrogallic some other acid, as

* For its method of preparation see Note 7.

much to preserve the solution as to develop the picture
uniformly; distilled water also must be employed, or at
least rain-water, well filtered : that which is obtained
during a storm contains ammonia or nitric acid, and is
therefore unfit for use.

Solution of pyrogallic acid for developing, is thus prepared :—A flask of about 1 pint capacity is obtained, and
made perfectly clean ; to this is adapted a funnel, furnished with a filter, as shown in Fig. 11. Upon this filter
throw 15 grs. of pyrogallic acid ; then 15 ozs. of distilled
water being measured off, and 1 oz. of crystallisable acetic
acid * being added thereto, the whole is well stirred with
a glass rod, and poured on to the filter containing the
pyrogallic acid, which latter dissolves as the liquid is
passing through. The filtration being complete the funnel
is removed, and the flask stopped lightly with a cork, and
set aside for use *in the dark.*

MM. Davanne and Girard have proposed the substi-
tution of citric for the acetic acid, in which case the
following formula may be employed :—

Distilled Water	12 fluid ounces.
Pyrogallic Acid	15 grains.
Citric Acid	15 ,,

Whichever formula is adopted, the method to be followed
to develop the picture is precisely the same ; and it is
important to note that the development should take place
as soon as possible after the plate is sensitised, and on no
account should a longer time than five minutes be allowed
to elapse between these operations.

The plate, on removal from the frame, is held by one

Fig. 58.

corner (the same by which it was held when
the collodion was poured over it); then
a sufficient quantity of the developing solution to cover the plate being placed in a
glass (Fig. 58), is poured by a continuous
operation over the collodion surface, and the
plate inclined alternately in every direction, in order that the whole film may be
covered without delay. Then suddenly turn the plate up-

* See Note 11.

right, holding it by the opposite corner to that from whence the liquid flows into the glass.

Fig. 59. Developing a Negative Proof.

The same solution is again poured on the plate, inclining it constantly from right to left, in order that the liquid may be kept in continual motion. The image gradually appears; and when it is considered sufficiently developed, the plate is washed by dipping it, at first very carefully, into a shallow porcelain tray, filled with water; it is then placed under a small stream of water, after which it is fixed. It is desirable, for washing the plate, to use a little apparatus represented in Fig. 60. It is simply a flask, filled with distilled water. The cork is perforated, so as to admit of the passage of two tubes. The highest—that which is to be blown through—passes

Fig. 60. Wash-bottle.

through the cork, and terminates directly under it; the other, on the contrary, rests in the water at the bottom of the bottle. The draughtsman has reversed this arrangement in the figure by mistake.

In using this wash-bottle, it is held by the neck while the operator blows through the higher tube, by which means the water is projected through the exit tube, and may be directed over the surface of the film, beginning at the centre, and passing gradually to the edges, thus avoiding disruption. Fig. 61 shows the operation better than any verbal description.

Fig. 61. Washing the Film.

In order to avoid taking the plate in the hand, an India-rubber plate-holder is sometimes employed with advantage. Fig. 62 shows a glass plate attached to such a plate-holder. To do this the plate is placed on a level table, and the holder grasped by the globe part in such a way as to force out the air; the edges are then moistened, in order to make them adhere more perfectly, and the pressure is withdrawn at the moment of bringing the plate-holder in contact with the glass by pressing the edges of it. Although the air has been forced out of the bottle, its elastic sides re-assume their shape, and the vacuum left inside causes the plate to adhere very firmly to

Fig. 62. Pneumatic Plate-holder.

the plate-holder. To detach it, it is only necessary to

compress the bottle as before. The picture having been developed and washed, is now fixed, an operation to be described in the next chapter.

The process of developing the picture, although apparently simple, is, nevertheless, a very delicate one, and one on which depends, in great measure, the success of the positive proof; it will not, therefore, be out of place to enter a little more into detail. The latent image produced by the action of light upon the iodide of silver, is brought out by the action of pyrogallic acid, aided by the nitrate of silver, with which the collodion film is impregnated. If too much pyrogallic developing solution is poured from the developing glass on to the plate, the details of the picture appear but slowly, and if too little there will not be enough to cover the plate completely; thus the medium lies between the two extremes.

Sufficient developing solution should be poured into the glass to cover the plate freely and completely, and no more ; and this should be poured on and off, and allowed to flow in various directions, in order to facilitate the mixture of the developer with the adherent nitrate of silver. By this method of manipulating it will be found that the development proceeds uniformly, and may be observed from time to time by examining the plate by transmitted light, as indicated in Fig. 59.

The development is continued according to the preceding directions until the required amount of intensity is obtained, and the plate is then rinsed with water to remove the developing fluid, so as to prevent its continued action.

At an ordinary temperature the picture is generally sufficiently developed in two minutes; but if it is necessary to force the development, it will seldom prove a successful picture. At a higher temperature it sometimes becomes necessary to stop the development in half a minute, to prevent the high lights from becoming too dense. Experience alone can give the knowledge necessary to determine exactly the when and how in this delicate operation.

As soon as the solution of pyrogallic acid covers the plate, the sky and the high lights of the picture begin to appear on the primrose-tinted film of iodide of silver; a few seconds after, the minor details make their appearance, becoming more and more vigorous, and, as often occurs

in a landscape, the sky darkens so much, that it is even
difficult to see the sun through it. This will give an idea
of the amount of intensity which the image acquires; and,
without experience, it is very difficult to indicate the
exact moment when the pyrogallic acid should be removed
by treatment with cold water.

The quantity of acetic acid added to the pyrogallic
also exercises an important influence on the results, and
in proportion as it is increased the more slowly the image
appears, and the more vigorous it is in character. When,
for example, a negative has to be developed taken from

Fig. 63. Positive Proof. From a Stereoscopic View by M. Ferrier.

a group of persons among whom there are some having
on white dresses, a large proportion of acetic acid is
necessary; while, on the contrary, a dull, grey, sombre
monument develops best with a small proportion of acetic
acid. The time and character of the development of a
picture also enables a conclusion to be arrived at, as to
whether the exposure in the camera has been too long or
too short.

In this latter case it is with difficulty that the sky and the other high lights of the picture are developed, and detail in the shadows is never obtained; even after long waiting the image will not appear, in which case it is necessary to take a fresh negative.

Fig. 64. Negative Proof, examined on the Side of the Glass not covered with Collodion.

If, on the contrary, the exposure has been too long, the negative is red and uniform; after fixing, there is an absence of vigour in the blacks, and there is a general fogging spread over every part of the image; the development also has taken place very rapidly. In each case another negative must be taken.

The most frequent error is over-exposure, arising from an inability to realise the possibility of taking a picture in so short a time.

For the purpose of illustration, suppose the monument represented at Fig. 63, representing a portion of the Acropolis at Athens, is to be reproduced.

The sun illuminates these statues with all its meridian

splendour; strong shadows are therefore a necessary consequence, which, in order to be correct, ought to be in
the negative.

Beginning with a negative which shall not have been
exposed in the camera more than five seconds, and successively increasing the time of exposure, five seconds in
a series of five negatives, the following will be the
result :—

The first negative will be very slow in developing;
after several minutes the sky will with difficulty be traced,
and some faint indication may be obtained of the parts
most strongly lighted.

In the second these portions will have acquired much
more vigour, and the shadows even will begin to yield
some indication of detail. The third negative will be a
good one; but the fourth will be, in every respect, all
that could be desired, possessing a due amount of vigour,
and presenting the aspect of Fig. 64.

The fifth negative, although exposed a longer time,
will have acquired a greyish colour. A good negative is
blue; but too long an exposure gives it a dull grey appearance. The shadows, instead of being transparent,
like glass itself, have upon them a sensible deposit, owing
to the too rapid development.

In proportion as the exposure is increased in any subsequent negative, the grey colour tends to become red;
there is a difficulty almost in distinguishing the shadows
and the sky; and, in one word, the negative loses all its
vigour.

The above are the principles upon which the photographer should proceed in the correction of his negatives.

It would be useless to pretend that an accurate judgment can be formed except by long experience. It is not
until after several months of assiduous research, and after
continual disappointment, that the beginner will acquire
that accuracy of judgment at a glance which will guide
him in the fugitive indications of the amount of exposure
necessary for a given subject; but, when once this knowledge is acquired, he will be amply rewarded by the
small number of defective negatives, as well as by the
amount of photographic knowledge acquired during the
apprenticeship.

It frequently happens that negatives are spotted, or covered with curious markings, either in parts or entirely. The four accompanying figures are some examples of

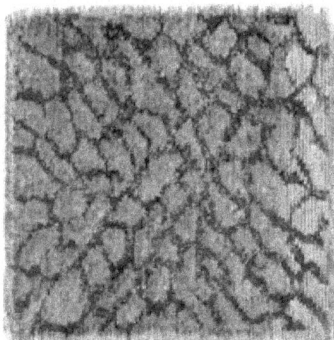
Fig. 65.

Fig. 66.

Fig. 67.

Fig. 68.

these spots, taken from M. de la Blanchere's work; the description of them, together with practical directions for their avoidance, will be found in Note 9 at the end of the volume.

CHAPTER X.

FIXING AND STRENGTHENING THE NEGATIVE IMAGE.

It has been before stated that when the image is sufficiently developed, the plate should be washed with
water, so that the pyrogallic acid may be well removed.
Should the film show any tendency to detach itself, this
operation must be performed with considerable care. The
small shutters may now be opened, so as to uncover the
whole of the yellow windows. The light in the dark
room is thus increased, although it still retains its yellow
tint.

The result thus obtained is a negative image, but still
partly obscured with iodide of silver, which is now to be
removed. For this purpose it must be plunged into a bath
composed as follows :—

> Water.. 40 ounces.
> Hyposulphite of Soda .... 8 ounces.

In about half a minute, if the bath is fresh, and some few
minutes if the bath has been much used, all the yellow
coating of iodide of silver is dissolved, and there remains
on the collodion film only the pure silver compound constituting the photographic image.

Hyposulphite of soda is a crystallised substance, very
soluble in water, and can be purchased at a very cheap
price.

The solution of hyposulphite of soda may be contained
in a vessel of zinc or gutta-percha. It does not stain the
hands; on the contrary, it will remove recent stains of
nitrate of silver.

Great care should be taken that the vessel containing
the bath of hyposulphite be kept separate, and that the
fingers be well washed before proceeding to take another
picture, because contact with this salt spoils both the
collodion and nitrate of silver bath.

In fixing the proof, it should not be allowed to remain

too long in the bath of hyposulphite, otherwise the fine details will be injured. The entire solution of the iodide of silver is easily perceived by examining the back of the glass. The iodide of silver being of a pale yellow colour is easily seen, and it is, however, necessary that the last traces of this colour be allowed to disappear. When this result is obtained, the glass plate should be taken out and plunged into a bath of cold water.

The solution of hyposulphite will serve a considerable time before it is exhausted, and then it is of so little cost that a fresh solution is easily made.

When the negative proof is immersed in the hyposulphite, the vessel containing it may be taken to the daylight, which will enable the operator better to see when the yellow coating of iodide of silver is quite removed. Ordinary daylight from this time has no action on the picture.

After the fixing is complete, it is of the greatest importance thoroughly to wash the collodion surface, especially if a great number of proofs are to be printed from it; this becomes more difficult when the coating has a great tendency to detach itself from the edges of the glass. Nevertheless, even under these circumstances, with a little attention it may be perfectly washed. When the collodion is good, and especially if the gun-cotton has been prepared according to the formula indicated in Note 3, there will not be the least fear of such an accident.

When the film appears liable to detach itself, the small apparatus, Fig. 60, which has already been described, is made use of. The current of water is carefully directed from the middle of the plate towards the edges, constantly varying the inclination in such a way that the water always flows from the centre towards the sides. By these means the coating is prevented, as far as practicable, from being totally removed. Care must also be taken when the glass is left to drain, as the coating wrinkles and bends by its own weight. This effect shows itself while the coating is yet quite wet. When the excess of water has disappeared, a little care will enable the coating to be drawn to its proper position by the aid of the finger.

To properly wash the coating the glass should be

placed, during five minutes at the least, in a large vessel
of zinc filled with water. If the coating separates in any
part the glass must be very gently removed, and the
detached parts arranged and brought together by a
very fine and light jet of water applied to the required
places. We have often in this way replaced a film upon
the glass after it has been entirely removed and torn at
the edges, and in spite of all obtained good results.

If the glass be not sufficiently washed after removal
from the hyposulphite, the film becomes sticky, and after
awhile disappears altogether, or, at least, becomes strongly
stained.

It is after removal from the hyposulphite of soda that
the proof can be best examined as to its quality. It ought
to be of a bluish colour; the sky and high lights a strong
black, and the deep shades nearly transparent.

If at this moment a good photographer judges that his
negative is wanting in vigour, he commences another;
but if this is impracticable, recourse must be had to *intensifying.* This operation must be done after the proof is
fixed and washed, and before it is dried.

Generally speaking, an intensified negative is worth
but little; and, to repeat, *a good operator will rather re-
commence another negative than intensify.* However, if
required, the best method is as follows :—

The glass plate, after the rinsing which follows the
fixing, is plunged into a porcelain vessel containing a
solution of bichloride of mercury in water.

Water... 10 ounces.
Bichloride of Mercury........................... to saturation.

(To saturate the water with bichloride of mercury, this
salt, after being powdered, is introduced into a bottle
filled with water, and well agitated.)

In a few minutes the coating assumes a milky-white
appearance. The plate is now removed and washed with
the greatest care, and then plunged into a solution of

Water .... 10 ounces,
Liq. Ammonia... 1 ounce;

when it immediately becomes darkened. On being taken
from this bath the plate is washed with water, and then
placed against the wall to drain and dry.

The best method of drying glass plates consists in placing them (Fig. 69) resting by their upper edge against the wall, and their lower edge upon a sheet of bibulous paper, which very quickly absorbs the excess of water.

If the plate is required to be quickly dried, the coated surface ought to be from the wall; and, on the contrary, if it is required to be slowly dried, the coated surface is placed towards the wall. In this latter case, the particles of dust do not so readily attach themselves to the surface of the picture.

Fig. 69.

The intensifying which we have described is very powerful; but it is capable of being regulated, by leaving the plate, for a few seconds only, in the bath of bichloride of mercury. The less time the plate is in the solution, the smaller the quantity of ammonia required to darken it. After a few years, a proof intensified by this method loses its vigour, and becomes even more feeble than it was before. Several other methods for intensifying have been proposed in various photographic works; none of them, however, can be recommended in preference to that described. At the same time, if a good proof be not obtained at the first attempt, by far the better plan will be to try again, as it is by this means alone that real perfection can be attained in the photographic art.

CHAPTER XI.

VARNISHING THE PICTURE.

WHEN the proof is fixed, washed, and dried, its surface exhibits by reflected light a fine metallic appearance; if the development has been carried too far, the coating appears covered with a metallic dust, which, however, can be partially removed with the aid of a very soft brush. By transmitted light the picture has an entirely different appearance, perfectly opposite to that exhibited by reflection.

As a collodion negative is required to furnish a considerable number of positive proofs upon paper, and for which purpose the collodion surface requires to be placed in contact with the sensitive paper, it becomes necessary to protect this coating by a sufficiently hard varnish, so that the plate can be easily handled without risk ; otherwise the soft coating would be very liable to injury.

There are two excellent varnishes suitable for the purpose—*amber varnish*, and the white gum, or *Soehnee varnish.*

The amber varnish is thus prepared. A quantity of yellow amber, broken into small pieces, is placed in a bottle so as to occupy about three-fourths of its capacity. A mixture of equal parts of chloroform and ether is now poured into the bottle, in such a quantity that the amber is hardly covered. After some few days the liquid contents of the bottle are poured upon a filter, and the pale yellow liquid which passes through is amber varnish.

The solid fragments of amber are allowed to remain in the bottle, to which can be again added the mixture of chloroform and ether, and a fresh quantity of varnish obtained. The same amber will serve for a dozen or more times, if care be taken to keep the bottle well closed.

To prepare the gum-lac, or *Soehnee varnish*, place in a bottle—

Gum-lac White........................	2½ ounces.
Alcohol	35 ounces.

The bottle is now well stopped and left several days, being well agitated at frequent intervals ; the liquid is then left to settle as much as possible, the clear portion poured off, and the remainder filtered. The whole of the liquid can be filtered ; but it is not recommended, as it passes very slowly through the paper. The colour of this varnish is of a pale yellow, less deep than the chloroform.

Instead of making this preparation in a bottle at the ordinary temperature, it will be found a more expeditious plan to employ a flask (containing the materials), placed in hot water. If the flask, thus kept heated, be agitated from time to time, the gum-lac can be dissolved in about half an hour. There is always a whitish deposit, which does not dissolve either in cold or hot alcohol ; this must be allowed to perfectly settle before the varnish is poured off for use.

The amber varnish is very easily applied ; the collodion surface of the plate, on which the picture is formed, is covered with it in the same way as the collodion was applied, the excess of liquid being received in the bottle. In a few seconds this varnish is dry. It is well, however, to expose the plate for an hour to the sun before being used for printing from.

It is more difficult to make use of the white gum-lac varnish, because the plate then requires to be warmed. For this purpose it is carefully exposed to the flame of a spirit lamp, or before a clear fire, taking care that the heat is equally applied by moving about the glass plate at some distance from the fire, or above the flame, if this be employed.

The temperature of the glass should be such that, when applied to the back of the hand, it can be borne without inconvenience. If too hot, the varnish flows with difficulty over the surface ; and if too cold, it does not dry sufficiently quick, and becomes *chilled* or partially opaque. This varnish is applied like collodion, the excess being received in the bottle, and when the surface appears dry, the heat is continued a short time longer, so as to cause a proper transparency and adhesion of the coating.

It is especially difficult to use the gum-lac varnish for plates of large size, for which, all things considered, it is generally best to employ the amber varnish. The negatives are less firmly varnished, and, in consequence, not able to yield so large a number of proofs ; but, on the other hand, it is much more easily applied.

It constantly happens that a number of transparent round spots appear on the negative, which require stopping out before the positive proofs are obtained from it, otherwise these white spots would form corresponding black ones on the prints, which cannot be removed. If,

on the contrary, the spots on the negative be black or
opaque, white spots are formed on the positive proof,
which are comparatively easy to be touched out by a
little water colour.

Touching out a negative requires very careful manipulation, and a properly arranged apparatus for the purpose
will be found very convenient, Fig. 70. The wooden top

Fig. 70. Apparatus for touching out Glass Proofs.

of a small table is removed, and a thick sheet of glass,
supported by an iron framing, is substituted. Below is
arranged a large sheet of white paper, stretched upon a
frame, or else a looking-glass, of which the inclination
can be varied by some such contrivance as shown in the
figure.

The light reflected from the looking-glass or paper,
enables the proof, which is laid on the glass-plate forming the top of the table, to be conveniently examined by
transmitted light.

The transparent spots are then touched out with a very
fine camel's hair pencil, and some dark colour mixed with

honey or oil.* Some persons touch out the spots with Indian ink, mixed with a small quantity of Prussian blue, before the plate is varnished. This plan, however, cannot be recommended, as the subsequent varnishing is apt to destroy the colour applied.

Light spots on paper proofs are easier to touch out with water colour of the same tint as the print, than glass negatives are: therefore, whenever circumstances allow a choice to be made, preference should be given to the former. For example, if there should be a white spot on the negative, with which nothing can be done, it will be better to make it a black one, which, by printing white on the paper proof, can be then easily tinted of the proper colour.

CHAPTER XII.

THE DRY COLLODION, COLLODIO-ALBUMEN, AND TANNIN PROCESSES.

1. *The Dry Collodion Process.*

ONE of the most unfortunate circumstances connected with the practice of photography on wet collodion, is the necessity, in travelling, for the photographer to employ a tent, which adds greatly to his luggage. This inconvenience is due to the fact that the operations of sensitising and developing must quickly follow each other, for, if too long a time elapses between them, the nitrate of silver crystallises on the surface of the glass, and causes an infinite number of small spots to be formed over the entire surface of the picture.

If the glass plate, covered with its sensitised coating

---

* M. de la Blanchère gives the following formula:—

Peroxide of Iron, or Rouge; Chromate of Lead, or Ivory Black	10 parts.
Gum Arabic, Saturated Solution	2 ,,
White Honey	2 ,,
Sugar Candy	1 part.

of collodion, be washed with distilled water when removed from the silver-bath, and left to dry, it loses nearly all its sensibility, and will give but very imperfect pictures.

M. Desprats and M. Dubosq have partially remedied this fault: the former, by adding half per cent. of resin; the latter, some few drops of amber varnish to the collodion; in whichever of the two methods it is prepared, it must be used exactly as ordinary collodion, with the exception that when the glass is removed from the silver-bath, it is to be well washed and dried in the dark. Plates thus prepared will preserve their sensibility for many days. The time of exposure in the camera ought to be tripled, and the development made with solution of gallic acid.

M. Dupuis recommends the adoption of the following process. Ordinary iodised collodion is employed, and the glass plate after being sensitised, is well washed in distilled water. A solution of *dextrine*, in ten times its weight of water, is left to settle, and then decanted, so as to be free from impurities. A requisite quantity of this solution is poured over the sensitised glass plate, recently washed, and while still moist, in such a manner, that it spreads evenly over its surface; it is then drained and dried.

The plates thus prepared will keep good for several days; but the time of exposure in the camera must be tripled, that is to say, if with wet collodion, 30 seconds of exposure is required, it will be requisite to give 90 with a plate prepared with dextrine.

Before developing the plate, it is washed with pure water. An apparatus, consisting of a flat-bottomed flask, &c., similar to that described page 57, is useful for this purpose. The picture is developed with pyrogallic acid.

A great number of other dry processes have been proposed; but two, above all others, have obtained the preference of the photographic public, as well in England as in France and Germany, these are the *collodio albumen process* of M. Taupenot, and the *tannin process* of Major Russel. These two processes will, therefore, be described in all their details.

2 The Collodio-albumen Process.

As by this process a considerable number of plates can be prepared in a few hours, which is very convenient, especially for obtaining stereoscopic positive proofs upon glass, which will presently be described. The following will be found an excellent method, and one which will enable a hundred of these plates to be properly prepared, in two operations of four hours' continuous work, it being understood that the glasses do not exceed the dimensions of 9 inches by 7 inches, and that they are all cleaned beforehand.

In the *dark room* are arranged two vessels of gutta-percha, one beside the other, both containing a silver-bath composed of—

> Water .... 35 ounces.
> Fused Nitrate of Silver......... 1050 grs.

One bath only may be employed, but with two the operation, as will be perceived, is much quicker.

Ordinary negative collodion may be used, but it will be found an improvement if the collodion is a little less iodised, such as is obtained by adding to the ordinary collodion—of which the formulæ has been given at page 22 —one third of its volume of a mixture of about two parts of ether and one of alcohol.

> Collodion, ordinary, (page 22)......... 3 ounces.
> Ether, 5 fluid drs. }
> Alcohol, 3 do....... } .... 1 ounce.

For the purpose of facilitating the description, the glass plates will be designated by the letters A, B, C, and the nitrate of silver baths by numbers 1 and 2.

A glass, A, is coated with collodion and plunged into the first silver-bath (No. 1). A second glass is now coated with collodion, and plunged into the second bath (No. 2).

By the side of the silver-baths should be placed a vessel of gutta-percha of considerable depth, and filled with filtered rain-water, or better, with distilled water. This bath ought to be as capacious as possible.

When the first glass, A, has been a sufficient time in the silver-bath, No. 1, which is ascertained by the dis-

appearance of the greasy veins which have been spoken
of at page 35, it is taken out and placed in the bath of
water contained in the vessel of gutta-percha.

A third glass, C, is then coated with collodion and
plunged into the silver-bath, No. 1, from which the first
glass was taken, and then the glass in the bath of water
is removed, and placed in a large wooden tub (Fig. 71)
filled with rain-water, and sufficiently large to hold from
eight to ten glasses, placed side by side.

Fig. 71. Tub for Washing Plates.

The second glass, B, is now taken from the silver-bath
and plunged into the bath of water in the gutta-percha
vessel.   It is left here while another glass is coated with
collodion and sensitised, and when that is finished the
second glass, B, is taken from the water, to be placed in
that contained in the large vessel Fig. 71.   The plate, C,
is removed to the bath of water, a fresh plate inserted,
and the same series of operations repeated for the number
of prepared plates required.

It will be understood that this method of preparation
being a continuous one, is necessarily very rapid.   When
the large vessel of water is filled with plates, they are to
be taken out and placed inclined against the wall, resting
at the bottom on some bibulous paper, and with their
prepared or collodionised surfaces towards the wall, as
described at page 67.

If the operation be conducted on a small scale, with
one glass at a time, then, when the first plate is taken
from the silver-bath and plunged into the water-bath,
another plate is coated with collodion and placed in the
nitrate-bath, and during the time it has to remain in it,

the first plate is removed to a fresh quantity of water contained in a trough of wood or gutta-percha.

Whichever method be employed, the result will be the same, provided that care be taken to remove the whole of the nitrate of silver, which covers the plates, with abundance of water; if the trough be small, the water must be often changed; but if it be of large size, this may be dispensed with.

When the whole of the plates are prepared, some solution of common salt is added to the water contained in the gutta-percha trough.

A white precipitate is formed of *chloride of silver*, which can be retained until of sufficient quantity, with other residuums, to be reduced into metallic silver by fusion in an earthen crucible with carbonate of soda.

Each glass plate should remain in the water at least ten minutes, and ought to be kept *upright*, the better to remove all the soluble salt of silver from the texture of the collodion coating.

The glasses, coated with collodion and washed as just described, must be covered *while still moist*, with iodised albumen, which is thus prepared:—

Fresh fowls' eggs are broken across the middle, the *whites* carefully separated, and then poured into a glazed earthen vessel, to which is added a quantity of iodide of potassium, equal to $7\frac{1}{2}$ grains to the white of each egg employed. Before adding the iodide of potassium to the white of egg, it should be dissolved in an equal weight of water ; for example, if ten, twenty, or fifty eggs be used, 75, 150, or 375 grains of iodide of potassium are required, dissolved in 75, 150, or 375 grains of water.

Fig. 72.   Fig. 73.

The whole is then beaten completely into froth by means of a bundle of twigs, represented at Fig. 72, or of tinned iron wire, mounted in a handle, Fig. 73. This froth is left to itself in a glazed earthen vessel, Fig. 74,

for twelve hours, when the greater portion is resolved
into clear albumen, which can be poured off into a wide-
mouthed bottle.   This iodised albumen is used to pour
over the collodionised glasses after they have been taken
from the trough, Fig. 71, and drained for one or two
minutes on a wooden shelf, Fig. 69.

Fig. 74. Albumen beaten into Froth.

M. Taupenot employed fermented albumen ; but it is
now not generally used or to be recommended.

Iodised albumen can be preserved for a considerable
time during winter, but in summer it is apt to decompose
very rapidly.   It should be kept in well-closed bottles
in a cool place.

During the time that some of the glasses require to
remain in the silver-bath, those which have been draining against the wall are to have a small quantity of the
albumen poured on each, and allowed to run over every
portion of the moist collodion surface ; the glass is then
held vertically, so as to allow the excess of liquid to flow
into a *separate bottle,* which, for the sake of distinction,
will be called B.

A fresh quantity of the iodised albumen is now
poured on the glass, and made to flow over every portion
of its surface, the excess being received back into the
bottle.

The glasses thus albumenised are left to dry, the
upper part of each leaning against the wall, and the
lower resting on some blotting-paper, as shown at Fig. 69.

The albumenised surface should be towards the wall, to avoid dust.

The reason for giving the plates two coatings of albumen is that the first application serves to remove the water which impregnates the collodion surface and allows the second quantity to give a perfectly uniform coating. The albumen contained in the bottle, B, can be used a considerable number of times for giving the first coating or until it becomes too diluted with the water derived from the plates, when a fresh quantity must be used; whereas, that employed for the second coating, on the contrary, can be used as often as required.

The glasses, when removed from the bath of water, should not be allowed to dry before being albumenised, otherwise the albumen is spread with difficulty on the surface, and stains are subsequently produced.

The albumen should not be applied sparingly to the glass, but at the same time it should not be allowed to flow over to the back of the plate; if this should happen, it is best removed, *when perfectly dry*, with some bibulous paper moistened with water.

Walking about the room when the albumenised glasses are being left to dry should be avoided as much as possible, so as to prevent any particles of dust, &c., settling on them. In about twelve hours they will be sufficiently dry, when they can be preserved for an indefinite time, if placed in a grooved box to protect them from damp and the direct light of day.

All the foregoing operations are to be done in the dark room, and when finished, the water in the large bath may be thrown away, but that in the smaller bath of gutta-percha should be retained, for the purpose of precipitating the silver it contains, as before described.

The following is a *resumé* of the first series of operations, in their proper order, that each glass undergoes :—

1. Cleaning.
2. Coating with collodion.
3. Sensitising.
4. Washing for a few moments in the first bath of water contained in the gutta-percha vessel.
5. Washing a second time in a bath, where it is left for several minutes.

6. Leaving to drain.
7. Coating with iodised albumen, which has been before used for removing the excess of water.
8. The immediate application of another coating of fresh iodised albumen.
9. Lastly, leaving it to drain and dry.

On the evening, or at most the day before requiring to use the plates for taking views, they must be submitted to a second series of operations, so as to render the iodised albumen surface sensitive to light.

A great number of albumenised glasses may by prepared at a time, because *they can be kept for an indefinite period;* whereas, no greater number of them should be sensitised than will be used in a very short time, as they then rapidly deteriorate by keeping.

For sensitising the glasses a bath of aceto-nitrate of silver is required, composed of—

Distilled Water	16 ounces
Glacial Acetic Acid	1 „
Fused Nitrate of Silver	1½ „

This bath requires to be filtered before being used, and a gutta-percha dish should be employed for it. After being used for a few weeks it becomes of a yellowish tint, that can, however, be removed by being shaken up with some kaolin. As a matter of precaution the kaolin may always remain at the bottom of the bottle, and the liquid decanted from it when used.

At the side of the bath of aceto-nitrate of silver should be placed another *of much larger size,* filled with filtered rain-water.

An albumenised glass is immersed with one quick movement into the aceto-nitrate of silver, and left there for fifteen seconds or more, it is then placed in the bath of rain-water, which is shaken a short time so as to remove the excess of nitrate of silver. This being done, it is taken out of the water and left to dry against the wall in the manner previously described. It is almost needless to say that the operation of sensitising must be done in the dark room with a yellow light. The glasses thus sensitised, when dry, are ready to receive the impression in the camera, after which they can be kept for a fortnight before being developed, although, as a rule, the shorter time that elapses between the sensitising and development the better will be the result. To obtain good proofs, not more than three days should intervene between the two operations. In summer especially, the time should be as short as possible; in winter it may be longer without so much risk of injury.

The sensitised glasses ought to be preserved in grooved boxes free from chinks or cracks through which the daylight might pass, and kept as dry as possible.

When required for use they are placed in the ordinary collodion frame for views. There are, however, some frames particularly devised for dry plates, which hold two glasses and have two slides, first one of the glasses is exposed to the light, and then the other, by simply turning round the frame.

The time of exposure for a collodio-albumenised plate is fully double or treble that required for ordinary collodion. Experience alone will guide the operator in this particular.

As the pictures are generally developed after returning from an excursion, it is an excellent plan, to ensure a good proof of any particular view, building, &c., to take two impressions, but with different times of exposure.

The collodio-albumen process is also particularly well adapted for obtaining transparent positives on glass. For this purpose the sensitised surface of the prepared plate is put in contact with the varnished surface of the negative, and placed in a close frame similar to that used for exposing the plate in the camera. By opening the sliding shutter the diffused light of day is allowed to act for three or four seconds, taking care that in arranging the glasses in the frame, the light passes through the negative before striking the sensitised plate. The plate is then taken into the dark room, and developed in the ordinary manner.

When this process is followed the positive picture obtained has a very good effect if placed before a window; it should be mounted with a plate of ground glass, the albumenised surface of one in contact with the ground surface of the other, and the two plates united by a border of black paper pasted round the edges.

The following is the method of developing the collodio-albumen plates :—

Place in a porcelain capsule 15 grains of *gallic acid*,
and pour over it $3\frac{1}{4}$ ounces of hot distilled water, and
mix them well with a glass rod ; when the gallic acid is
dissolved, or nearly so, add 13 ounces of cold water, and
filter the whole into a bottle for use.

This liquid is poured into a *porcelain* dish to about
the depth of an inch, and to each quantity of three ounces
employed is to be added one-fourth of a drachm of the
following solution :—

Distilled Water...	35 ounces.
Fused Nitrate of Silver.............................	230 grains.
Glacial Acetic Acid	9 minims.

And care should be taken that the whole be thoroughly
mingled together, *otherwise stains will be formed on the
surface of the negative.*

The exposed albumenised plate is to be quickly
plunged into this liquid, the coated surface upwards,
and raised up and down for several times by means of a
hook so that the fluid flows well over its surface. It is
necessary that the porcelain dish should be adapted for
the purpose, and with a flat bottom so that the glass
may be perfectly *immersed* in the solution.

At the end of the first hour or two the sky and other
parts highly illuminated will hardly have made their
appearance, but in the succeeding two hours the proof
usually comes out with extreme vigour. It is always
well to watch this operation so as to stop it so soon as
the development is complete, which, however, sometimes
takes as long as twelve hours. The colder the weather
the longer the time required. In winter it is preferable,
and sometimes necessary, to develop in a warmed
apartment.

Very often in about three or four hours the development of the image can be accelerated by renewing the
gallic acid and doubling the dose of nitrate of silver,
that is to say, to each 3 ounces of solution of gallic acid
add a half instead of a quarter drachm of the aceto-
nitrate of silver; but if this is done the dish must be kept
agitated, otherwise the particles of reduced silver which

are formed would attach themselves to the picture and
entirely spoil it.

To succeed well, two conditions are indispensable—a
warm room, about 68° Fahr., and a slow development.
It is also very important that the *gallic acid should not
be allowed to become coloured and muddy;* if this should
happen, the glass must be washed, so as to remove the
thick and decomposed gallic acid from its surface, and
placed in another dish with some fresh solution of gallic
acid, to which has been added a less quantity of aceto-
nitrate of silver than was employed at first.

This last observation is most important, and the operation is one that cannot be too often repeated; with a
proper attention to the time of exposure and consequent
regulation of the doses of gallic acid, the picture should
be well and perfectly developed in about four hours.

To finish off the proof, it is washed with water when
taken from the gallic acid bath, and fixed with hyposulphite of soda in the manner described for the ordinary
wet process—a final washing completes the picture. It
is useless to varnish the plate, as the albumen itself
forms an excellent preservative coating.

As the beauty of a collodio-albumen proof greatly
depends upon the proper time of exposure in the camera,
it may be useful to give some of the indications of a too
short or too lengthened action of light.

If the exposure has been too short, the picture does
not come out well even after twelve hours of development and repeated changing of the gallic acid solution,
the sky alone is distinctly marked, all the other parts of
the proof have ‘a general grey tint, without the least
vigour. Nothing can remedy this fault. If, on the contrary, the exposure has been too long, the sky shows
itself very quickly, as well as the other light parts of the
picture. In about an hour the proof exhibits a considerable amount of intensity in all its details. If the
development be now stopped, a passable picture may be
obtained; but if it be continued, the whole of the image
becomes grey, and the sky, although apparently very
dark, because viewed by reflection from a white ground,
is, nevertheless, also grey. Over exposure is likewise indicated by a red tint, which the proof takes after being fixed.

When the time of exposure has been correct, the sky begins to show itself in about an hour, and very gradually increases in intensity, and becomes of an absolute black, even when viewed as a transparency after fixing. The whole of the details are shown with great vigour, while the shadows are perfectly transparent.

If the sky should not be sufficiently dark, and it is not possible to take another proof, recourse may be had to stopping out with a little Indian ink mixed with honey and water, and a soft camel's hair pencil. It is, however, next to impossible to do this operation in such a manner that some of the details on the horizon are not injured.

3. *The Tannin Process.*

This dry process derives its name from the use of tannin—a bitter principle obtained from gall-nuts—as a preservative agent. To Major Russell is due the credit of having introduced it.

The glass to be prepared should be cleaned with great care, particularly from any greasy substances. This is conveniently done with a mixture of Tripoli powder, spirits of wine, and solution of ammonia. A tuft of cotton is dipped into this mixture and rubbed over its surface for a minute or so; then well rinsed in water and rubbed dry with a clean cloth.

The glass, just before being used, should be wiped with a perfectly dry and warm cloth, and then coated with the following solution :—

> Nelson's Patent Gelatine..................... 20 grains.
> Distilled Water .... 10 ounces.
> Alcohol .... ½ ounce.

Dissolve and filter; this solution will keep good for a considerable time.

This gelatine solution is applied to the glass in the same way as ordinary collodion, taking care that the whole of the surface is covered, and that the back of the plate be not soiled. The superfluous liquid is received back into the bottle, and the plate set to dry, as shown at Fig. 69; when well drained, remove the accumulation of fluid very carefully from the lower edge of the plate by a piece of blotting-paper drawn along it. When the surface is dry, warm gently by the fire, and retain for use in a grooved box. As plates thus coated will keep good any length of time, any required number may be prepared, taking care that the backs of them are quite free from stains of gelatine.

The gelatinised glass is now coated with *old iodised collodion* in the usual manner, taking particular care that the *whole* surface of the plate be covered; it is then immersed in the silver-bath employed for the wet collodion process (page 34), and allowed to remain in it from three to five minutes.

Remove the sensitive plate from the bath, and wash it freely under a water-tap for about a minute, it will then be ready to receive the preservative solution, composed as follows:—

Tannin .... 60 grains.
Distilled Water .... 4 ounces.

Filter through paper, and measure out two separate portions according to the size of the plate to be prepared, allowing about two drachms in each quantity for a stereoscopic plate. The first portion of tannin solution is poured over the washed coating of the sensitised plate two or three times, so as to remove the water adherent to it, then the other quantity is poured on and off, and the plate placed on end on a piece of blotting-paper, and allowed to dry in a perfectly dark and warm place.

After exposure in the camera, which averages from one to three minutes on a favourable day, and from four to eight minutes in dull weather, the picture is to be developed, for which purpose the following solutions are required:—

No. 1. { Pyrogallic Acid...................... 72 grains.
{ Alcohol .... 1 ounce.

Dissolve and keep in a stoppered bottle.

No. 2. { Nitrate of Silver............ 20 grains.
{ Citric Acid .... 20 ,,
{ Distilled Water .... 1 ounce.

Dissolve and filter should any white or other precipitate be formed. To three ounces of distilled water add half a drachm of No. 1, and if the plate to be developed be a stereoscopic size, take three drachms of this solution and add to it from ten to twenty minims of No. 2; this forms the developing fluid.

The exposed plate is first moistened with distilled water, which must be done quickly and evenly, otherwise stains are produced, and then the developing fluid poured over its surface and kept slightly in motion. The development must be carefully watched, and if in a short time the sky comes out strongly, but is not followed by the other details of the object, the plate was not long enough exposed, and the developing fluid must be poured back into the measure, and say ten minims of No. 1 added, so as to increase the quantity of pyrogallic acid. If the whole of the picture, however, appears to come out at once, a few drops of No. 2 is to be added, so as to increase the density of the sky.

When the picture is properly developed, it is fixed with a solution of hyposulphite of soda, washed and varnished as described for the wet collodion process.

CHAPTER XIII.

POSITIVE COLLODION PROCESS.

The positive collodion process is very similar in its general details to the negative, and will generally be found much easier of manipulation, from its not requiring the like amount of accuracy.

Preparation of the Collodion.

Iodide of Cadmium, or Ammonium	15 grains.
Pyroxyline	15 ,,
Ether	3½ ounces.
Alcohol...	1½ ,,

The pyroxyline and the iodide of cadmium are first introduced into a dry flask or bottle, and the alcohol poured upon them, and the mixture shaken violently for about a minute; the ether is then added and the contents further agitated, and finally set aside all night. The clear supernatant portion may now be carefully poured off from the white deposit at the bottom of the bottle, or instead of decanting, it may be preferable to use a small syphon bottle as shown at Fig. 23, the action of which has already been explained.

The positive collodion thus prepared should, like the negative, be preserved in the dark in well corked or stoppered bottles.

Should the collodion produce a fogged image, or in other words, if the plate becomes covered with a film of reduced silver, underneath which there is obviously a very good picture, this defect may be remedied by adding to it a few drops of the following solution :—

Alcohol .... 3½ ounces.
Iodine .... 150 grains.

A small quantity of this liquid is added to the collodion, so as to impart a pale amber colour thereto.

Patent glass plates are not required for direct collodion positives, and the reason is very simple. When negatives are taken they are afterwards applied to sheets of sensitised paper with some degree of pressure, in order to obtain the positive impression, on which account it is necessary that the plate should be perfectly flat and free from striæ, otherwise the striæ would be reproduced upon the paper, and the glass very liable to fracture from the pressure. Such is not the case, however, with direct positives ; here the purity and flatness of the glass is by no means so important, and ordinary flatted crown may therefore be used. Some photographers employ glass of a deep red or purple colour, in which case it is unnecessary to varnish the plate.

Whatever kind of glass plate be adopted, it should be chosen as flat as possible, otherwise there will be difficulty in getting them into the camera back ; but the colour in no way affects the beauty of the image, any more than bubbles or other mechanical surface defects.

It has been previously explained, when describing the negative collodion process, how the plate is cleaned, and in what manner the collodion is poured on the glass ; no further remarks are therefore needed on the subject beyond reminding the reader that the collodion should be poured upon the plate in the dark room, in order to be ready as soon as the film is set to plunge it into the bath, which is composed as follows—

Distilled Water	35 ounces.
Fused Nitrate of Silver..................	2¼ ,,

Filter.   Prepare also the following solution—

Water	3¼ ounces.
Acetate of Ammonia	30 grains.

And, after filtration, pour it into the bottle which contains the above solution of nitrate of silver. A crystalline precipitate is produced which requires several days to deposit before the bath will be ready for use. It may, however, be used a few hours after mixing by resorting to filtration; but it is always better after a few days' keeping.

The bottle containing the bath should have a funnel and filter adapted to it, through which the solution ought to be always filtered after use, so as to be ready for each day's work, the same filter serving almost indefinitely.

The mixture of acetate of silver with the nitrate not only produces increased rapidity, but tends to maintain it. It has, moreover, the advantage of preventing fogging; but to this end it is necessary to be always kept in the dark.

The details of manipulation, as before mentioned, are the same as have been described for the negative process, except that the exposure is reduced about three-fourths and the development effected by protosulphate of iron instead of pyrogallic acid.

A saturated solution of protosulphate of iron is prepared by pouring half a gallon of boiling water on two pounds of *green vitriol*, or ordinary protosulphate of iron, stirring well together ; allow it to cool and then filter.

This saturated solution is, of course, much too concentrated for use ; the *developing solution* is therefore prepared from it in the following manner—

Water	18	ounces.
Saturated Solution of Sulphate of Iron	8	,,
Ordinary Acetic Acid of Commerce	1¾	,,
Ordinary Alcohol	1	,,
Sulphuric Acid....................................	1	,,

This solution should be contained in a vertical bath with a dipper, and the plate immersed therein for fifteen seconds after exposure, which will be found amply sufficient to develop the picture in all cases.

On withdrawal from the bath the image ought to possess very little intensity, if otherwise it might be regarded as an indication that the proof would not be a satisfactory one. However this may be, the plate is now well washed with water, fixed with cyanide of potassium, and dried as has been described under the negative process.

It is only after fixing that a correct idea can be formed as to whether the exposure has been correct or not, and practice alone will enable the operator to determine the point. If any trace of fogging should become apparent, a few drops of the alcoholic solution of iodine may be added to the collodion, the formula for which has been already given.

It may not be out of place here to give a few explanatory remarks in order that the reader may fully comprehend the essential difference between a positive and negative collodion picture.

A negative proof being intended to possess varying degrees of opacity to transmitted light, it is necessary that the film should have in it a sufficient *quantity of material* to suffer various amounts of decomposition in the process of development; to this end, not only is a thick collodion employed, but also one fully charged with iodide.

With a positive the case is entirely different, the film is extremely thin, for unless it were so the whole would be patchy, owing to the light having penetrated the film and effaced the delicate details by reflection. It is, therefore, of the highest importance to use the thinnest possible collodion, which yields, on leaving the bath, a film only faintly opalescent.

For fixing, cyanide of potassium is a more appropriate agent than hyposulphite of soda, as it yields a more agreeable tone. It requires, however, to be used with great care on account of its highly poisonous character. The removal of the unaltered iodide by its means may be effected either by immersion or by pouring the solution over the plate. The fixing solution is prepared by mixing together—

Water	35 ounces.
Cyanide of Potassium	1½ to 3 ounces.

Longer contact of the cyanide than is absolutely necessary should be avoided, as the delicate detail is apt to suffer, and when the iodide is removed, which may be ascertained by the transparency which the plate acquires, it should be freely washed with water, dried, and varnished.

The varnish, if the picture be taken on a colourless glass, must be black on the plain side, and colourless or "crystal varnish" on the film side; when purple glass, however, is used, the black varnish may be dispensed with.

A solution of gum arabic is sometimes poured over the film side of the plate, instead of varnish. Some operators omit all kind of protection; this is, however, a mistake, as pictures which are not covered either with gum arabic or crystal varnish, very quickly assume a tarnished and disagreeable aspect.

To communicate an extra amount of whiteness to direct positives they are sometimes treated with bichloride of mercury after the final washing; the general tone, however, is usually sufficiently cold, and therefore its employment cannot be recommended in an artistic point of view.

The method of operating is as follows :—Cold distilled water is allowed to take up as much bichloride of mercury as it will, and, the solution being filtered, the positive is immersed therein until it assumes an appearance as white as snow, which will be in about one minute, after which it is washed and varnished in the usual way. These proofs should be preserved from the direct action of the sun's rays, otherwise they become very much weakened in a few months.

The different methods of mounting in *passe-partouts*, &c., involve operations of so strictly mechanical a nature that they need not be entered into in this place, as the means of carrying them out are soon acquired by practice.

Glass positives are very easy to take, although some practice is necessary to attain to any great degree of excellence. The pictures should have a mirror-like aspect, and the blacks extremely pure. Sometimes a certain amount of fogging takes place, marring the brilliancy of the picture; this may be removed by washing the picture while still wet, immediately after fixing, with

an alcoholic solution of iodine of 15 grains to the ounce.
The length of time must be determined by the extent
of the fogging, and can only be ascertained by practice.
When it is considered that the action of the iodine has
been sufficiently prolonged, it is washed off, and the plate
again immersed in the cyanide fixing bath, and finally
washed and dried.

CHAPTER XIV.

THE STEREOSCOPE.

This instrument shows in relief, as one picture, two dis-
similar views of the same object, mounted together either
on plate or paper. The following is an explanation of
this singular phenomenon. If, for example, a square-
based pyramid be examined with the left eye, the eye
being placed in the axis of this pyramid, it is very
evident that it will present the appearance of a square
intersected by its two diagonals, and a drawing of it
would, in fact, be represented by such a figure. But if
the left eye be now closed, and the right eye opened, it is
equally clear that one side of the pyramid will be more
visible than the other, and that in proportion to its
height.

Now, if this same pyramid be viewed with both eyes
at the same time, the two different images combine upon
the retina in such a way as to produce the appearance of
a single solid pyramid, in perfect relief, or, in other words,
the effect is *stereoscopic*.

If the pyramid, instead of being in relief, were hollow,
and the left eye, as before, placed on a line with its axis,
a square figure intersected by its two diagonals will be
seen; but immediately upon opening the right eye, the
effect will be exactly the reverse of what occurred in the
case of the figure in relief. In the former case, it was
the side of the pyramid to the right which impinged
mostly on the right eye, but in the case of the hollow

figure, it would be, on the contrary, the left eye which receives the larger impression, and in consequence thereof, the effect upon the two eyes would be different, and instead of producing the sensation of solidity or relief, the combined images produce the appearance of hollowness, or in other words, are *pseudoscopic.*

It is upon this principle that the stereoscope is founded, the two views employed together are each taken by a camera, as near as possible under the same conditions as naturally presented to each eye. The prismatic glasses of the instrument superpose the two views in such a way that but one image is percieved by the two eyes, and that in relief.

To Professor Wheatstone is due the credit for the elaboration of this theory and the application of photography thereto; also the invention of one modification of the instrument. Mirrors were employed in its construction, and it is known as the reflecting stereoscope. Sir David Brewster replaced the mirrors by prisms, and it is this improvement which has rendered it so popular and portable.

The relief of near objects is easily explained, by the large angle formed by the visual rays brought to each eye; but as this angle becomes more and more diminished in proportion to the distance of the objects, it becomes at last very difficult to appreciate relief, and in consequence, the situation of large masses can only be judged of by comparison with those which are near.

Fig. 75. Prismatic Stereoscope.

The original stereoscope was composed of small half lenses, which turned in brass tubes in such a way as completely to superpose the two views; but now the small lenses are replaced by large square prisms, let into the upper portion of the stereoscope, and this not only affords a larger field of view, but also makes the instrument more easy to manage.

Fig. 75 represents a prismatic stereoscope. The slides to be viewed are introduced through a groove at the bottom ; it is furnished with a centre partition, which assists in the superposition of the two images. The upper portion is shown detached, or open, the better to exhibit its general construction.

For the best instruments the two prisms are achromatic, or else replaced by two achromatic lenses, by which their definition is rendered as perfect as possible. A very convenient form of stereoscope for showing a number of slides is now constructed, whereby as many as fifty pictures can be viewed consecutively, by merely turning a small button. This arrangement consists of a stereoscope fixed at the top of a deep box containing the views, which are attached to a kind of endless chain, on moving which the pictures are presented as required.

There are many other forms of stereoscope, the whole of which, however, with the exception, perhaps, of the reflecting stereoscope for viewing large pictures, being constructed on the general principle and arrangements of the foregoing, any further description would be unnecessary.

Stereoscopic views are always of small size, except

Fig. 76. Twin-lens Camera with Double Back.   Fig. 77.

those before mentioned for the reflecting stereoscope, and are taken on glass and paper ; those on glass are viewed as transparencies, but the paper slides are examined by reflected light, for which purpose there is usually attached to the stereoscope some such contrivance as is represented at Fig. 75, like a small door in front.

The most simple method of obtaining stereoscopic pictures certainly consists in taking the two proofs at the same time, with the apparatus as represented in Fig. 76. It is composed, 1st, of two double achromatic combinations of exactly the same focal length ; 2nd, of an ordinary expanding camera, furnished, however, with a back, having two shutters (Fig. 77).

The following is the method of operating :—

Whenever the two lenses are not exactly to the same focus, it is necessary to begin by focussing a near object, such, for instance, as a plaster bust. When a perfectly sharp image is obtained upon the ground glass with each of the lenses, the rack and pinion adjustment must not be again touched, but the focussing of any subsequent figure must be accomplished by drawing in or out the expanding or inner body of the camera.

This, however, will not suffice, if the distance of the bust from the camera has not been so regulated as to be that which will be most suitable for a portrait. For this apparatus will serve only for portraits, and that, too, for portraits taken at a very short distance ; otherwise, the proofs will be wanting in relief.

As would be inferred from the shape of the back, a glass is chosen much longer than it is wide, upon which the two negatives are taken at the same time. The prints obtained from these negatives are reversed, that is to say, when each positive is cut for the purpose of being mounted upon card, it will be necessary to paste the left-hand picture on the right side of the card, and the right-hand print on the left of the card.

When, with this apparatus, it is desired to take views, a diaphragm or stop must be inserted between the two lenses, but the two proofs can, however, no longer be taken at the same time, for the distance between the two lenses is not sufficient to give the proper amount of stereoscopic relief for distant objects.

A board with moveable guides is therefore placed upon the camera-stand, as represented at Fig. 81, and two cross lines are drawn upon the ground glass of the camera. One marks the centre on one side ; the second, that on the other. The camera is then pointed in such a way that the same object in the landscape comes on the centre of the cross lines, for without this the proof would not be properly centred.

The two negatives then are taken separately, and require transposing when they are mounted on the card. But this reversing or transposition may be, however, accomplished in the instrument itself, so effectively, that when the negative is placed in the stereoscope, the objects appear in relief.

For this purpose, suppose the apparatus, Fig. 76, is in the position as represented in Fig. 81. Instead of taking with the right-hand lens the right-hand negative, and with the left-hand lens the left-hand negative, let the operations be reversed. Take the *left* picture with the *right* lens, and *vice versâ*—the *right* picture with the *left* lens.

When it is desired to take landscapes only, the camera represented in Fig. 78 is generally used. It is an ordinary quarter-plate camera, which is converted into a stereoscopic camera by the addition of three pieces of apparatus, two of which are represented in Figs. 79 and 80.

In Fig. 79 will be seen a grooved piece of wood, with two rebates and a catch in the middle of the upper one, and a ground glass sliding between them; this is stopped in the middle by the catch before mentioned. This piece of wood is screwed to an ordinary camera-back, the door of which is reversed.

Two diagonal lines are drawn upon the ground glass. When it is desired to take a stereoscopic view, the board with two moveable rebates, as represented in Fig. 81, is brought into use.

The following is the method of proceeding. The camera, Fig. 78, is first placed in one of its positions, and directed towards the view—then in the second position, and directed towards the same point. These two views will

Fig. 78. Camera for stereoscopic Views.

appear upon the ground glass exactly the same, but they are not so in reality.

The ground glass being withdrawn, the camera is placed on the right hand of the operator, who looks at the view which is before him. Replacing the focussing glass by the camera-back, which is introduced on the right side of the piece of wood (Fig. 79), it is obvious that it is

Fig. 79. Frame with Ground Glass.

Fig. 80. Frame for Sensitive Plate.

that part of the glass which is on the left in the back which will be first uncovered, as is shown in Fig. 80. The slide being drawn, and the time of exposure having expired, the door is closed. Fig. 78 shows the position of the camera for the *first picture*, that is to say, the right-hand one.

In the second position, that of the left, the camera-back is made to slide between the two rebates until it is checked by the brass catch; the second picture is then taken.

These directions seem complicated, but in practice they prove very simple. The principle being once understood, it is easy to follow the description.

There remains a third method of operating, which is still more difficult. When it is desired to take views of animated objects or groups, the necessary interval which elapses between the two positions of the camera, just described, renders it inapplicable for the purpose, as in all probability the objects must have moved, and the two views would not be alike. Under these circumstances two quarter-plate cameras are used, furnished with lenses whose focal lengths are exactly equal, and the operation is conducted as is shown in Fig. 81. Great care must be taken to open and close the two shutters at the same time, or at least very quickly one after the other, as it is obvious that if one plate be longer exposed

than the other, the development must take place unequally, and the two proofs will be unequal also.

Fig. 81. Position of the two Cameras for Instantaneous Views.

In order that the two negatives should be exactly
alike, the two plates should be collodionised at the
same time, and immersed one after the other in the same
nitrate-bath, from which they should be simultaneously
withdrawn.

In reference to the development, a dish is used with
a glass bottom and wooden sides, or of solid glass,
and the two plates are placed therein, one by the side of
the other, in order that both may be developed with
the same pyrogallic acid. This is the only way to
obtain proofs of *nearly* equal intensity, for it is, indeed,
quite a chance that they should be exactly alike; a
little difference will not, however, materially affect the
result in printing.

Positive proofs should be reversed: the one which was
taken on the right should be on the left in the card, and
in order that this may always be done correctly, it is
advisable to mark each of the negatives when taken
separately.

The apparatus for *cartes de visite*, at page 39, is, in
fact, nothing more than the apparatus, Fig. 76, with
a back like Fig. 80. It consists of a camera to which
four lenses of equal focal length have been adapted,
furnished with a sliding back. From this arrangement
it follows that when the lenses are arranged as shown in
Fig. 76, there will be eight negatives taken on each plate,
and that each pair will be stereoscopic. The cost of the
apparatus may, however, be materially lessened by using
a whole-plate camera fitted with a pair of twin-lenses,

as in Fig. 76, and arranged so as to slide together on its rising front; in this way four *carte de visite* pictures may be obtained on one plate. A number of negatives are taken on one plate, because the printing of an equal number of proofs separately would be much more costly and troublesome.

It remains to consider the proper angular distance which should separate the two cameras represented in Fig. 81, so as to produce the desired result.

If several square-based pyramids, of various heights, be placed successively on the same spot, and examined with one eye from a fixed point over and slightly to the side of the pyramids, it will be perceived that each time a pyramid is changed, that the relative position of the apex alters, and that this displacement is greater the taller the pyramid viewed, or in other words, the relief is more visible. If these pyramids be now viewed again with one eye, but this time not on one side, but directly on a line with their axes, they will, although all differing in height, nevertheless present the same appearance of a square cut with its diagonals.

Thus, each time that the height of the pyramid varies, the visual angle becomes greater, and the relief is consequently more striking.

But suppose for a moment that it is desired to reproduce the pyramid, what angle should be employed? should it be the angle of vision, or a larger one?

In view of what has been said above, it is evident that if a larger angle be used, the pyramid will appear in the stereoscope higher than it really is, and as the photographic apparatus does not furnish any standard of comparison by means of an object of known height placed side by side, a false idea will be conveyed of the true size of the pyramid.

When the angle is too great, the object is distorted. This defect is, so to speak, very general in all reproductions of statues, portraits, &c. The relief in these cases is, with very few exceptions, greatly exaggerated. And thus it happens that when a statue is reproduced in this way, the head often seems so much in advance, that it becomes really monstrous.

It should always be borne in mind when objects are to

be reproduced which are of known size and proportions, such as statues, portraits, animals, &c., that if a large angle be employed the image will be distorted.

For views and landscapes an excessive relief may be obtained without detriment to the picture, on the contrary, with a decided improvement in effect, at the same time attention must be had to the character of the view. If, for example, the foreground of the picture be very near the operator, a small angle (about 2°) must be used, otherwise the two pictures will not become superposed in the stereoscope. If, on the contrary, the view offers but little difference between the various planes, advantage may be taken of this, and a larger angle, 4° for example, may be employed.

The stereoscopic angle is reckoned from the nearest point of the view to be taken; this point forming the apex of the triangle, and the distance between the two cameras being the base, or rather the arc, of the angle.

When the operator has chosen his point of sight, the eye should estimate the difference between the cameras and the nearest point in the view to be taken, which will determine whether an angle of 2° or 4° be preferable. The following table will, to a certain extent, serve as a guide for beginners to judge the value of an angle, that of 2° being selected as an example:—

Distance of the Object from the Cameras. Yards.	Distance between the Cameras. Inches.
1	1·26
2	2·52
3	3·78
4	5·04
5	6·30
6	7·56
7	8·82
8	10·08
9	11·34
10	12·60
15	18·90
20	25·20
25	31·68
30	37·80
35	44·28
40	50·40

CHAPTER XV.

PRINTING POSITIVE PROOFS.

THEORETICALLY, the printing process is one of the greatest simplicity :—One side of a sheet of paper is first imbued with a solution of common salt, and dried; it is then treated with a solution of nitrate of silver. A white substance, called chloride of silver, is formed in the texture of the paper, which has the singular property of becoming rapidly darkened on exposure to the sun ; it is on this account that when it is prepared the operation should be performed in the absence of daylight, and the sensitive paper preserved in a dark box.

It is evident that if paper so prepared be placed behind a negative, and exposed to daylight, a positive image will be produced on the paper. And, in order to its preservation, it will be necessary to remove the whole of the chloride of silver unacted on by light, without affecting those portions which constitute the image ; this is readily accomplished by soaking the proof for a quarter of an hour in a solution of hyposulphite of soda, and then washing most copiously in water to remove every trace of this fixing agent from the texture of the paper.

The proof thus completed is the final result of all the operations which have been previously described.

The foregoing description gives the general principles of the printing process ; a successful result, however, can only be attained by the exercise of the greatest care and attention to many important details, which will be treated of under the following heads :—

1. *Salting the paper*, that is, treating the paper with a solution of common salt, or chloride of ammonium.
2. *Sensitising the paper*, by floating on a solution of nitrate of silver.
3. *Exposing the prepared paper to light*, and the apparatus used for this purpose.
4. *Fixing the image formed by light, and imparting an agreeable tone.*
5. *Mounting the proofs*, that is to say, giving the final touches before delivering them to the public.

1. *Salting the Paper.*

Suitable paper for positive proofs is now manufactured expressly for the purpose, and can readily be obtained both in England and on the continent. It is not, however, every sheet in the ream as obtained from the *manufacturers* that ought to be employed. Those that are uneven in texture, and blemished with black and other spots, must be rejected.

While selecting paper it is important to touch only the edges of the sheets, because, however dry the fingers may be, they always leave a slight imprint which, sooner or later, will produce a stain.

It is also necessary to ascertain the *right* side of the paper to which the chemicals should be applied; this may be done by examining the paper by reflected light, and is that side most uniform in its character and freest from lines. Each sheet, as selected, should have the wrong side distinguished by a pencil mark.

Positive paper should be moderately thick. A ream should weigh about 22 lbs., the size being 22 inches by 17 inches; but, generally speaking, for albumenised paper, the ream should not weigh less than 24 lbs., especially if it be required highly glazed; while for ordinary salted paper a weight of 18 lbs. to the ream will be found sufficient. The paper may be salted with various chlorides, such as the chlorides of barium, strontium, potassium, ammonium, &c.; and although it was once thought that each salt imparted a characteristic tint to the proof, a contrary opinion is now entertained; however, *chloride of sodium*, or *common salt*, which is always to be obtained in a fair state of purity, is now generally used.

The salt is dissolved in distilled, or rain-water, in the proportion of 12 grains to the ounce, and the solution filtered into a dish large enough to hold a whole sheet of paper, and measuring, therefore, about 24 inches by 19 inches. The preparation of *whole* sheets of paper at one time will generally be found most advantageous, and they are afterwards easily cut to any smaller sizes that may be required.

The paper is spread out or floated on this solution in the following manner :—The opposite ends of the sheet

F 2

are held, and the paper bent as represented in Fig. 82, and the middle of the paper being brought into contact with the liquid, the two ends are regularly lowered until the whole of one side just floats on the surface. The side of the paper previously marked with a pencil should be uppermost.

The paper ought to remain on this bath five minutes in winter and three minutes in summer, at the end of which

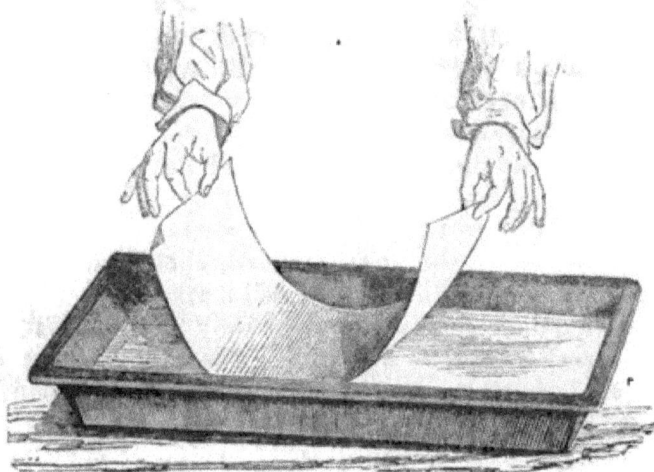

Fig. 82. Salting the Paper.

time it is withdrawn by means of a pair of horn or box-wood forceps and held for a little while over the dish, to drain; it is then attached to two clips, or other means of suspension, and hung on a line to dry (Fig. 84).

If the operator, or rather amateur, should not be often performing this operation it may be found more conve-nient to use smaller sheets, and to suspend them in the manner indicated in Fig. 83; whichever method of sus-pension, however, is adopted, a small piece of white bibulous paper should be attached to the lower corner or corners, in order to aid the draining of the salt solution.

Paper thus salted is preserved in a portfolio, and, as it does not undergo any alteration by keeping, it may be prepared in large quantities at a time.

Albumenised paper yields positive proofs of great

delicacy and fine colour. It is, however, a little more difficult to prepare.

White of egg is beaten into a snow-white froth, adding, for each 10 eggs, 150 grains of salt, *reduced to fine*

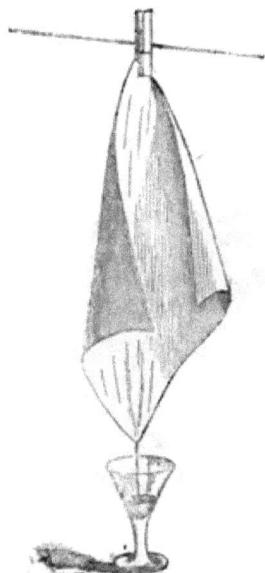

Fig. 83.   *Fig. 84.*

Hanging the Paper to Dry.

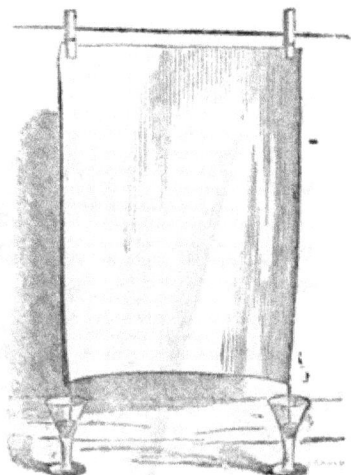

powder; then, when the albumen has accumulated by being allowed to remain at rest (see Fig. 74, and pages 75, 76), it is poured into a large flat porcelain tray, and used in the same manner as the ordinary salting-bath, except that the viscous nature of the albumen requires greater care in floating.

It is very difficult to albumenise large sheets of paper. The first operation should be the removal of bubbles by means of a card, and this must be done very effectually, because wherever there is a bubble the liquid does not touch the paper, and the consequence is a white spot in the print.

It is advisable to keep the paper which it is intended to albumenise in a damp place for a few days before using, otherwise it gets into folds on the liquid, and bubbles are sure to form. When, however, the paper is damp it spreads with the greatest facility on the albumen-bath, and no fear need be entertained as to the formation of the fatal bubbles. These precautions are specially applicable to large sheets, the small ones being comparatively easy of manipulation.

The large sheets are very difficult to suspend. An excellent plan is to attach the sheets, on being withdrawn

Fig. 85. Drying of Large Sheets.

from the bath, to frames of cardboard by the aid of pins, as shown in Fig. 85. In this way the tearing of the sheet, and its frequent falling to the ground, is avoided.

There is not much purely albumenised paper used; that which has been prepared with a mixture of albumen and water being preferred, on account of the more moderate brilliancy given to the sheets. From 10 to 40 per cent. of water may be added to the pure albumen; but the more water, the less brilliant the paper.

For portraiture a paper prepared with pure albumen

should not be used, on account of its excessive brilliancy; but rather plain salted paper, or paper prepared by floating on a mixture of equal volumes of albumen and water containing 4 per cent. of salt. Albumenised paper should be preserved in tin or zinc boxes, closed from contact with a damp atmosphere, for if it absorbs moisture it becomes rapidly changed, especially in summer time. Whether the amateur prepares his own paper, or whether he buys it, there are always some bad sheets. Sometimes there are great lines arising from the draining of the albumen towards one point, sometimes the gloss is greater at one place than another, without the reason being very obvious.

2. Sensitising the Paper.

This operation should be performed in the dark, or at all events in a moderate light. Daylight does not act by any means so energetically on positive paper as on collodion; nevertheless, it is advisable to have a room specially arranged for the purpose.

To this end the windows should be furnished with yellow blinds or frames, which will admit a large body of light, having no action on the prepared paper.

Paper is sensitised by floating it for three or four minutes on a silver-bath of 20 per cent.

Water	20 ounces.
Recrystallised Nitrate of Silver	4 ,,

This solution is poured into a perfectly clean dish, and the paper floated thereon, in the manner shown at Fig. 82.

If the proportion of nitrate of silver be much reduced the proofs will be wanting in vigour; from which circumstance, as often as 100 sheets of paper, 9 inches by 7, have been sensitised, half an ounce of nitrate of silver should be added to the bath.

In order to avoid soiling the fingers, and also the stains which would be produced by the contact of the clips, it is best to turn up a corner of the sheet (Fig. 86), which prevents its absorbing any of the nitrate of silver, and may, therefore, be taken hold of for hanging up the sheet. Instead of hanging the sheets of paper by hooks (Fig. 87), they can be more conveniently suspended by some little articles, supplied with india-rubber springs, known as American Clips (Fig. 88).

It is only necessary to press on the ends of these clips in order to open them, and to withdraw the pressure after

 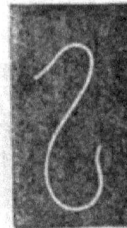

Fig. 86.   Fig. 87.

inserting the paper, which is then held with the greatest tenacity. Figs. 83 and 84 show how these improved clips are used.

Small pieces of white blotting-paper should be attached to the corners of the sheet from which the solution is draining, and the drops of nitrate of silver may be collected in test-glasses (Figs. 83 and 84), although the drops so collected serve only for transformation into chloride of silver.

The dry sensitised sheets shoul d be us ed within a day or two after preparation, unless preserved in *a box with chloride of calcium,* to be presently described.

Fig. 88.

The silver-bath may, of course, be used until it is exhausted, remembering always to add the half ounce of nitrate of silver for every 100 sheets 9 inches by 7, as before mentioned. It must, however, be filtered every time it is desired to sensitise a fresh batch of paper.

A special bath should be used for albumenised paper, to which should be added one-fortieth of its weight of *kaolin,* which should always be kept at the bottom of the bottle, shaking it up from time to time, and regularly allowing it to settle. This kaolin maintains the bath in a colourless condition, for without it the solution would acquire a deep yellow colour which would be communicated to the proofs.

When a number of sheets have been sensitised, they

are placed in portfolios, one on the other, handling them only by the edges, in order to avoid contact of the fingers, which produces stains. They should be used within three days.

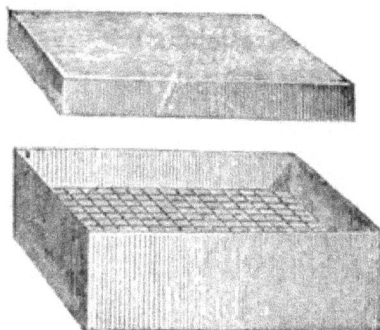

Fig. 89. Apparatus for preserving Sensitised Paper.

If it be desired to preserve sensitised paper for some months, an apparatus must be employed consisting of a box with an air-tight cover, and containing some *dried chloride of calcium.* Figs. 89 and 90 represent two forms of the apparatus. Fig. 89 consists of a square zinc or tin box, about six inches deep, furnished with a false bottom, on the top of which is placed the sensitised paper, and underneath a metal or porcelain basin filled with pieces of *dried chloride of calcium;* the whole is closed by a well-fitting lid, which can be rendered air-tight by pasting a strip of paper around the joining.

The other form, Fig. 90, is a very convenient one for travelling. The outer cylinder B and cover A are made of zinc or tin plate; the inner core E D is formed of canvas or metal gauze, and is filled with pieces of dried chloride of calcium. A sheet of blotting-paper is first rolled round this core, and afterwards the sheets of sensitised paper; it is then put into the cylinder, the lid adjusted, and a band of vulcanised india-rubber placed over the joining, as shown at C, which renders the whole air-tight.

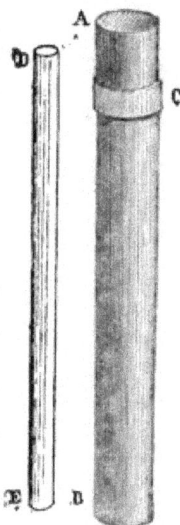

Fig. 90.

Whenever any of the sheets of paper are required to

be taken from either form of apparatus, it should be closed
again as quickly as possible, because the chloride of calcium
very rapidly attracts moisture from the air; from which
circumstance its peculiar action of keeping perfectly dry
the air in the interior of the boxes depends.

When the chloride of calcium has been some time in
use, it becomes covered with moisture, and eventually
converted into a liquid, and its action destroyed. So soon
as any moisture is perceived, it ought to be placed in a
moderately hot oven, in a porcelain or iron vessel, until
perfectly dry, when its properties will be entirely restored.
Thus the same quantity of chloride of calcium can be used
over and over again for any length of time required.

3. *Exposure of the Prepared Paper to Light.*

A printing frame is procured, furnished with screws
for large plates, and with
springs for small nega-
tives. The instrument
(Fig. 91) consists of a
simple wooden frame,
at the bottom of which
is a strong piece of
plate glass, which should
be always well cleaned
on both sides before
using. The negative is
placed on this glass with
the plain side down-

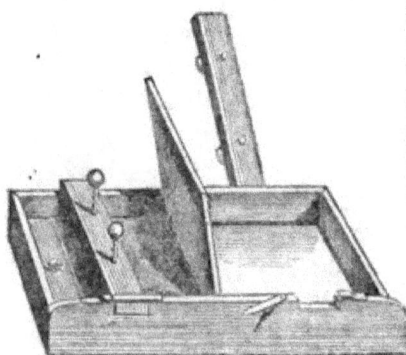

Fig. 91. Printing Frame.

wards, while the sensitised side of the paper is brought
into contact with the collodion film, which is, of course,
uppermost.

In order to maintain perfect contact between the sensitised paper and the negative, the printing-frame is furnished with a hinged board lined with cloth or felt, and
kept in its place by transverse bars with springs or screws.
If it be desired to examine the progress of the printing,
one of the transverse bars is removed, the hinged board
is opened, and one half of the paper bent back, as shown
in Fig. 92. This operation should, of course, not be performed in the direct sun-light, but in the shade.

A consideration of the following list of successive changes, through which the paper passes, will guide the operator in judging when a print has been sufficiently exposed.

Fig. 92. How to examine the Action of Light on the Paper.

The paper turns—
1. Very pale blue.
2. Pale blue.
3. Clear bluish purple.
4. Deep purple.
5. Black.
6. Metallic greyish black.
7. Olive, or greenish bronze.

It is not sufficient that the picture when examined, as shown in Fig. 92, should *seem* to be printed deep enough, because the fixing agent removes a great deal of the depth of the colour; on this account, and in order that the finished print should possess the proper depth of tone, it is necessary that the tint imparted by exposure in the printing frame should be much deeper than is required for the finished picture.

The time, then, for withdrawing the paper from the action of the sun or light, depends solely on the method of fixing, and is best learned by experience.

Pressure-frames should be made with the greatest care, for if, for example, the back for pressing the paper against the glass should not be quite flat, the paper will in some places be away from the negative, while in other parts the contact will be complete, and the resulting positive image will bear evidence of unequal sharpness, thus marring the beauty of the picture.

Professional photographers, who generally have a great many negatives to print from, arrange the pressure-frames on a framework, so constructed as to admit of the angle of inclination being altered in such a way as to receive the solar rays perpendicularly.

The more vigorous and fully developed the negative, the more brilliant and satisfactory is the resulting positive proof. But if the negative be a weak one, it is possible, notwithstanding, to produce a passable print from it by

prolonging the action of the solar rays. To this end the frame is exposed not to the direct rays of the sun, but to diffused daylight. Or better still, the printing frame may be covered, as shown in Fig. 94, by a light frame, upon which is stretched a piece of thick white paper. When it is desired to produce portraits

Fig. 93.   Vignette Glasses for Gra-duated Backgrounds.

with a white background, glasses with yellow borders and white centres are employed, called vignette glasses, as shown in Fig. 93.

It will be understood that the yellow portion of these glasses prevents the passage of the chemical rays, and as the action of light on the chloride of silver takes place only through the colourless portion of the glass, that is to say, in the centre, it follows that from an ordinary negative a picture may be obtained with a white background.

When the background of a negative is not sufficiently developed, or is defective through any other cause, it becomes very difficult to remedy. Some persons paint in the background, others cut out the figure from a positive print and paste the matrix, or background portion, on the

Fig. 94. Moveable Framework for Printing Positives.

negative; but in either case the outline of the picture is always hard and disagreeable. A better plan is to cover the pressure frame with the vignette glass, Fig. 93, taking care, of course, to place the oval in such a position that only the bust shall be printed, while the defective background is as much as possible masked.

The following is the method adopted when it is desired to produce the effect of clouds in a proof; firstly, a negative is taken with the ordinary exposure, in which the sky will be perfectly black; secondly, another negative is taken with such a short exposure that nothing appears when the developing solution is poured on but the sky. With this short exposure the effect of clouds is very successfully obtained.

A positive being printed from the first negative will have, of course, a white sky; this being placed under the second, or cloud negative, the foreground of which is covered with cotton wool in such a way that light can pass only through that portion of the glass bearing the image of the clouds. The outline of the objects in the landscape is covered with cotton wool, the fibres of which have been well pulled out, in order that the hardness which would be otherwise produced may be moderated and toned down. An exposure of a few minutes will be found sufficient to produce the desired effect.

To produce a similar effect artificially is even still more easy. In taking the negative it is arranged in the exposure that the sky shall not be too deep, and the desired cloud effects are painted, not on the film, but on the other side of the glass. It should be remembered, however, that the effect produced in the positive proof will be the reverse of what is painted on the glass.

A recently varnished negative should not be immediately printed from under the direct rays of the sun, for however good the varnish employed may be, it invariably sticks a little when freshly applied. Instead of exposing to direct sunlight, use diffused daylight, or better still, cover the pressure-frame with a case or covering of white paper.

Paper should be made perfectly dry after sensitising, and before being placed in contact with the negative, otherwise the moisture in the fibres of the paper will be volatilised by the sun's heat, and condense on the negative, which will soon become blackened by the nitrate of silver of the paper. When this is discovered in time, the spots so produced may be removed by cyanide of potassium; the operation, however, requires great dexterity and judgment.

Negatives should never be allowed to remain all night in contact with paper at all damp, as the depression of temperature induces a condensation of moisture on the surface of the glass, and brings about the same untoward result as occurs under the circumstances detailed in the preceding paragraph.

Unusual precautions are necessary in the use of albumenised paper, which easily sticks to the negative, and shows stains wherever it is touched. It should be handled only by the extreme edges.

It is very important that the woodwork of pressure-frames should be well put together, so that no warping shall occur through changes of temperature or differences in the hygrometric condition of the atmosphere; otherwise the prints will not be equally sharp over the entire surface, owing to the unequal contact between the paper and glass induced by the warping. Moreover, the negatives are very likely to become cracked from the same cause.

A great many sheets of positive paper may be prepared at a time, on the sole condition of keeping them in a box with chloride of calcium, as described at page 105. This box, however, ought not to be opened oftener than is absolutely necessary, but enough paper should be withdrawn at one time to serve for two days' consumption.

Before placing the negative in the pressure-frame, the plate-glass should be carefully wiped, in order to avoid the accumulation of dust on the proof.

When the paper is placed in the frame the screws must be tightly adjusted, so as to secure it against the slightest movement when the hinged back is opened to examine the progress of the printing.

The frame itself should be a little larger than the hinged back, in order that the latter may open without difficulty. If it be so tight as to require on effort to open it, there will almost invariably occur a displacement of the negative, and a consequent doubling of the outline in the proof.

4. *Toning and Fixing the Print.*

The process of *toning* is employed to impart a pleasing tint to the proof, and the object of *fixing* is to remove the unaltered chloride of silver by immersion in a solution of hyposulphite of soda.

In order to avoid constantly touching the proof during

these operations, horn or boxwood forceps are employed,
or glass rods bent as shown in Figs. 95 and 96.

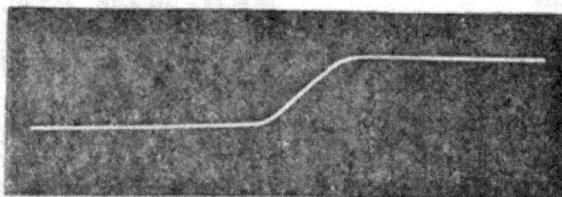

Fig. 95.

Many formulæ have been given for toning and fixing 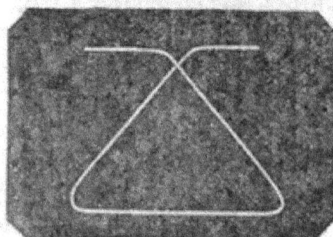 the prints, in some of which the two operations are conducted *simultaneously ;* in others the fixing is done before the toning, and in a third the toning is effected before the fixing.

Fig. 96. The latter process, however, is the one by which the
best results and the greatest permanency are obtained.
It consists of four operations : 1. Washing the print, to
remove the free nitrate of silver adhering to the surface,
and which has remained unchanged by light.   2. Toning
in a weak solution of chloride of gold, to which a small
quantity of carbonate of soda has been added.   3.
Washing again to get rid of the chloride of gold.   4.
Fixing in a solution of hyposulphite of soda.

The chloride of calcium box (Fig. 89) serves as well
for the preservation of proofs which have been exposed
to light as for sensitised paper.   As it is desirable to fix
a large number of proofs at one time, it is recommended
that the operator should at the outset provide himself
with three, at least, of the chloride of calcium boxes ;
one for preserving the white chloride, another for the
paper ready for immediate use, and the third for paper
which has been exposed under a negative.

The room set apart for the production of positive
proofs should be on the ground-floor, and the panes of
glass darkened with yellow paper ; it should be furnished
with several large wooden troughs for washing, and an
assortment of porcelain or brown stoneware trays for

hyposulphite solutions, &c. There should also be an abundant supply of water.

The proofs being taken either from the pressure-frame or the chloride of calcium box, are plunged one after the other into a porcelain pan containing rain-water, which is changed for every hundred prints. This bath should be at least one foot deep.

By far the greater quantity of the nitrate of silver withdrawn from the sensitising bath still remains on the surface of the paper after exposure to light. The bath of rain-water dissolves the nitrate of silver, and when the water is thoroughly charged therewith, it may be treated as described at page 75, by throwing in an excess of common salt, which produces a precipitate of chloride of silver.

There should not be more than ten prints washed at a time in this bath, and special care should be taken to prevent any hyposulphite of soda from falling therein, on which account the bath may be advantageously covered with a plate of glass, and the prints touched only with boxwood forceps.

When the prints have been for about ten minutes in the water, they are taken out one at a time and immersed in the toning-bath. To prepare the toning-bath, make the two following standard solutions :—

No. 1. { Bicarbonate Soda .... 2 drachms.
        { Water .... 32 ounces.

No. 2. { Chloride Gold .... 30 grains.
        { Distilled Water .... 32 ounces.

Take one quart of water and dissolve in it two drachms of common salt ; then add one ounce of solution No. 1 and one ounce of No. 2. This toning-bath should be kept about ten minutes before being used.

The first prints which are immersed in the bath after the addition of the chloride of gold tone rapidly ; but in proportion as the gold is taken up, the toning proceeds more slowly, when more of solution No. 2 will have to be added. To accelerate the toning in *winter*, the solution may be slightly warmed by setting the dish on a sheet of plate iron heated with a spirit lamp.

The chloride of sodium or common salt is added to the

bath to transform into chloride the nitrate of silver which has not been washed out, and which would otherwise decompose the chloride of gold.

The prints being put into the toning-bath will first become red, and will then pass successively through all the intermediate tints between red and black. The toning can be stopped at any stage by immersing the print in a tray of clean water, which should be kept at hand for the purpose.

The real colour is only seen after immersion in the hyposulphite fixing solution, and it is at this stage only that the operator can judge to what tint he should tone his pictures, in order that they should *finally* possess the desired colour. As a general rule the bluish black changes to a pure black in the hyposulphite; less toning than is required to give the bluish black produces prints more or less brown; if the toning is pushed further than the bluish black, the prints will be ashy coloured and flat when dry.

Prints which have to be toned black must be printed deeper than those which have to be toned purple or brown, for the toning-bath has also a bleaching action.

Prints on albumenised paper require for toning a larger quantity of gold than prints on ordinary salted paper.

The toning-bath may be used over and over again, by adding from time to time more gold and bicarbonate of soda; but, as hyposulphite of soda is liable to be accidentally introduced into it, it is advisable, where a large number of prints are produced, to make a new one every day.

In order to judge of the colour of the proof it should be withdrawn from the gold-bath by means of a boxwood forceps, as, on account of the yellow colour of the bath, it is difficult to form an accurate estimate while the print remains therein. The same remark may be applied to the colour of the panes of glass in the printing room. It has been said that they should be covered with yellow paper, but this applies, of course, only to those panes which illuminate that part of the room in which the prints are either withdrawn from the pressure-frame or placed therein, or where the first washing with plain water takes place. All the other operations may be safely conducted by daylight, although it will be found quite as well to preserve the whole of the workshop from the direct action of the solar rays by means of blinds.

The prints, as soon as they are taken out of the toning-bath, are immersed in water; they are then washed several times, and put into the fixing-bath, which is composed as follows :—

Hyposulphite of Soda .... 8 ounces.
Water................................. 1 pint.

The fixing will generally require from ten to fifteen minutes; but it is easy to understand that the longer the bath has been in use, the more will its capacity for dissolving chloride of silver become exhausted, and the longer will the print require immersion, so that, one pound of hyposulphite of soda will not fix more than one hundred prints, $8\frac{1}{2}$ in. $\times$ $6\frac{1}{2}$ in.; but as this salt is very cheap, to use it liberally will make but a fractional increase in the cost of the prints.

The hyposulphite bath should, then, be often changed, and when out of use it may be kept for the purpose of extracting the silver.

The proof which comes from the last bath is now rinsed in a wooden trough filled with water, to remove the excess of adherent hyposulphite; then allowed to soak for two hours in a second trough, and for three hours in a third. There should be an extra trough or dish for every twenty-five prints of $8\frac{1}{2}$ in. $\times$ $6\frac{1}{2}$ in. When the prints are allowed to remain longer in the water than is specified above, they lose their beautiful colour and become yellow.

It is sometimes desired to finish the washing of a print somewhat hurriedly, and yet to ensure the complete removal of every trace of hyposulphite of soda; when this is the case, the proof is placed on a glass plate of suitable size, and moved about under a stream of water from a tap falling from a height of about six feet. The motion imparted to the glass should be so regulated as to allow the water to fall on every part of the proof in succession, without tearing it or causing its separation from the glass. Five minutes of this energetic washing removes the hyposulphite more effectually than five hours soaking in the troughs of water, although it is obvious that the consumption of water is much greater. A badly fixed or imperfectly washed print undergoes spontaneous alteration after a few weeks.

The following may be received as practical rules to avoid the fading of prints.

As silver forms, with hyposulphite of soda, a double salt, slightly soluble in a feeble excess of hyposulphite of soda, but on the contrary very soluble in a concentrated solution,—and as this double hyposulphite of silver and soda decomposes with great facility,—it is very important that concentrated solutions should be used, in order that the double salt shall be dissolved as soon as formed.

A bath such as that of which the formula has been given, must not be used too long, and for the same reason too many proofs must not be immersed therein at one time, especially if there be not sufficient space between each proof to allow the bath to exert its full solvent action.

All the formulæ which prescribe the addition of an alkali or an acid to the bath should be unhesitatingly rejected. The chloride of gold of commerce is sometimes very acid, in which case it may be replaced, if desired, by a white salt known by the name of Sel d'Or de Fordos et Gelis; *this must be very much diluted before mixing with the hyposulphite of soda.*

If the proof, through insufficient washing, contains hyposulphite of soda, it will have, when dry, a sweet taste, and will fade in a very short time.

Finally, all fermented or fermentable substances for mounting photographs, should be rejected. Nothing but *recently prepared* starch-paste should be used, and the proof should be allowed to dry very rapidly when mounted, in order that no acid principles may be developed by fermentation.

In rainy climates in winter—in England, for example —it is sometimes necessary to expose the paper for a whole day to the light before a positive can be obtained. Albumenised paper being much more rapid than simple salted paper, will therefore be preferred, although even this will sometimes fail to produce a positive in a reasonable time, in which case recourse may be had to the following expedient.

Plain paper being cut to the proper dimensions, and the reverse side marked with a pencil, is allowed to float for about one minute on a bath composed of 300 grains of iodide of potassium dissolved in 35 ounces of water. This bath serves until it is exhausted, and the paper when dry and placed in a portfolio will keep a long time. The paper so prepared is rendered sensitive to light by floating (in the dark, of course) on a bath of nitrate of silver for one minute. The bath is composed of—

Water	3½	ounces.
Fused Nitrate Silver	30.	grains.
Glacial Acetic Acid	300	,,

The paper is withdrawn from the bath, well drained, and rapidly dried between folds of blotting-paper, in order to remove the excess of solution, and prevent its touching the back of the paper. This operation is repeated with a second and even a third sheet of blotting-paper. It is necessary that the sensitised sheet should be damp, but not to such a degree as to stain the negative, upon which it should be pressed *very lightly*.

The exposure to diffused daylight varies between five and fifteen seconds ; a long exposure yields a flat picture, deficient in brilliancy—in a short one the sky only appears.

The paper is carefully removed from the negative, and placed on a sheet of glass with the sensitised side uppermost. A solution is then made of 15 grains of gallic acid in 35 ounces of warm water, to which is added 3½ drachms of glacial acetic acid. This is filtered and spread very quickly by means of a brush or glass rod over the sensitised side of the sheet. The image appears very rapidly, and as soon as the desired effect is arrived at, the paper is plunged into water to arrest the action of the developer.

After washing, it is toned with chloride of gold, as described at page 113, and it is then washed for two hours in a large trough of water.

5. Mounting the Proofs.

Whatever process may have been adopted in the production of the positive print, it has finally to be pasted on Bristol board, and rolled.

Nothing is more simple than the mounting of a proof. It is cut into any suitable or desired form, and the back covered with starch-paste. It is then spread on the Bristol board, a piece of blotting-paper laid over it, and the whole rubbed with the hand in every direction, in order to ensure complete contact between the print and the board. This being done, it is allowed to dry in the air, and subsequently rolled in a lithographic or other press, by which it acquires a finishing gloss, which is entirely absent in unrolled prints.

Poirier's press, represented in Fig. 97, is very applicable for this purpose. It is composed of a steel roller, E, the axle blocks of which are moveable in a vertical direction; double levers are employed to make the necessary adjustment, and the power is applied at b b. It is sufficient to move one only of these heads in order to move the other, there being an arrangement of cog-wheels for this purpose. However, for delicate adjustments, the power is best applied to each screw-head separately, which, in turning, depresses two screws attached to the axle-blocks, these latter sliding between two edges of cast-iron.

The other part of the apparatus consists of a plate of planed polished steel, a a, or of a very flat lithographic stone, which receives its backward and forward motion from a toothed wheel, moved by a still smaller one, which in its turn is moved by a long lever or handle.

Upon this steel plate two very smooth and specially prepared cards are placed, between which the proof to be rolled is laid. The screws being adjusted, and the motion imparted, the steel plate moves on, turning the cylinder at the same time. An enormous pressure is thus brought to bear uniformly over the whole surface of the card, and after two or three movements to and fro, the screws b b are slightly tightened, and the rolling repeated.

The proof thus becomes pressed into the cardboard,

and acquires a splendid gloss. To mount stereoscopic
views and *cartes de visite,* they should be cut by means of
a steel or glass mould and a good knife.

It is a good plan to paste a great many proofs on one
card, in order to effect the glazing by one operation.

A beautiful gloss can be imparted to stereoscopic and
carte de visite proofs by varnishing them. This varnish is
made as follows :—One ounce of white lac is dissolved

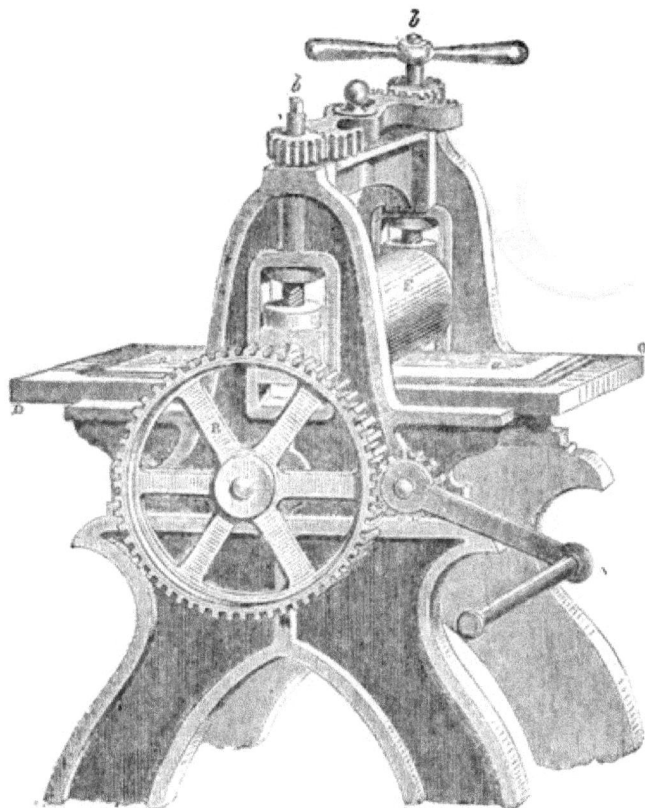

Fig. 97. Rolling Press.

in ten ounces of warm alcohol, and after allowing the
bottle to stand for several weeks, the clear portion is
decanted for use.

This varnish is then applied in a similar manner to
that adopted by French polishers in polishing cabinet

work; it has been introduced into commerce under the name of Crystal Enamel.

The picture having been mounted, is sized with a warm solution made by dissolving 10 grains of Swinborne's gelatine in 1 ounce of water, and either hot-pressed or burnished with an agate burnisher. It is then coated with the crystal enamel, applied in the following manner : —A tampion of cotton-wool saturated with the liquid is wrapped in a piece of clean white calico rag, the outer surface of which is touched with a drop or two of linseed oil : this is gently and evenly applied with a circular motion over the whole surface to be enamelled, until the picture becomes brilliant. It is, lastly, finished off by applying, in the same manner, alcohol and linseed oil.

NOTES.

NOTE 1 (*page* 4).

APPEARANCE OF NEGATIVES.

The difference of the reverse appearance of an ordinary negative on glass or paper, and the direct one exhibited by a metal plate, is simply apparent and not real, and arises solely from the mirror-like appearance in those parts of the silver plate not acted on by light. And this explains the reason why it is necessary to give to these (daguerreotype) pictures a particular position in relation to the angle at which the light strikes them. At other angles, when, for example, the light is directly reflected from the surface of the metal, the image appears as a negative.

NOTE 2 (*page* 16).

DISTILLATION OF ETHER.

The object aimed at in the distillation of a liquid is the separation of any solid substance which it holds in solution, or any liquid of a different constitution with which it may be mixed. In the instance under consideration, it is necessary to separate the ether, not only from the chloride of calcium upon which it has been dried, but also from the alcohol and water which it contains.

The boiling point of ether being about 96° Fahr., and of alcohol 174° Fahr., and water 212° Fahr., it follows, that if a mixture of the three be submitted to the action of heat, the ether will be almost completely volatilised before any sensible evaporation of the alcohol or water has taken place. A tube thermometer, having its stem passed through the cork, and its bulb so arranged as not to come into contact with the liquid, will indicate a temperature of about 96°, rising higher in proportion as more alcohol and water become evaporated. To secure pure ether, the temperature should not rise higher than 104°.

G

The first portions which come over should not be used, as they generally serve only to clean the condensing tube, and, consequently, contain impurities. From two pints of ether, therefore, the first ounce and a half which comes over should be rejected.

————

NOTE 3 (*page* 18).

PREPARATION OF PYROXYLINE.

The cotton should be chosen free from defects and any contaminating organic matter. A mixture is then made of

Sulphuric Acid (Sp. Gr. 1·8) 38 fluid ounces.
Nitric Acid..... (Sp. Gr. 1·4) 19 do. do.

This is stirred with a glass rod, and if examined by a thermometer will be found to indicate a temperature of 176° Fahr.; the operator should therefore wait until it cools down to 140° Fahr. before plunging in the cotton.

The quantity of cotton to use is about 1,050 grains, which is added to the acids, about one-fourth or one-fifth part at a time, squeezing it with the glass rods in order to force out the air imprisoned between the fibres. When all the cotton is immersed, the containing vessel is covered with a plate to keep in the nitrous vapours, and at the end of ten minutes the cotton is withdrawn, and copiously drenched with water, as before described. The pyroxyline thus obtained is less soluble in ether and alcohol than that obtained by the ordinary method; but it is especially useful when great tenacity of film is required.

————

NOTE 4 (*page* 19).

EMPLOYMENT OF THE IODIDES AND BROMIDES.

Besides iodide and bromide of cadmium, a great number of other iodides have been used, among which may be mentioned those of potassium, sodium, ammonium, zinc, &c.

Many photographers confine themselves to the use of the iodides of potassium and ammonium, but lately iodide of cadmium has come generally into use.

More recently has been proposed, especially for copying pictures, a collodion, containing iodide, bromide, and chloride of ethylamine. The following is the formula which has given the best results :—

Alcohol	1¼ ounce.
Ether	2½ ,,
Pyroxyline	15 grains.
Iodide of Ethylamine	1·2 ,,
Bromide ditto	0·4 ,,
Chloride ditto	0·2 ,,

Although iodide of ethylamine is not found in commerce, it is very easy to prepare. It is an organic iodide, containing nitrogen, and the elements of alcohol, but is nevertheless more stable than iodine of ammonium, and yields pictures of remarkable delicacy.

NOTE 5 (*page* 34).

NITRATE OF SILVER STAINS.

Nitrate of silver is reduced by contact with all organic substances, and as finely-divided metallic silver is black, it follows that this substance blackens everything it touches, as every photographer knows. Many methods have been proposed for the removal of these spots. The following is the best.

The hands become inevitably stained while conducting the manipulations involved in the sensitising and development of the plates, but it is not until some hours have elapsed that the stains, at first scarcely visible, deepen to any great extent; and it is only when a great number of proofs have been taken that the method about to be proposed is required or advisable.

It consists simply in well washing the hands in a saturated solution of hyposulphite of soda, kept expressly for that purpose. Two or three minutes' contact are found sufficient to remove every trace of the stain. Instead of hyposulphite of soda, iodide of potassium may be used in the same way. After the hyposulphite, it is advisable to wash the hands with soap, and with powdered and sifted pumice stone.

If the stains are very old, it is better to allow them to wear away by time; however, as they sometimes are *obliged* to be removed, in that case, a mixture of cyanide of potassium and iodine applied to the fingers by a brush, or otherwise, will, if aided by the use of a lump of pumice stone, rapidly restore the hands to their normal condition, after which they should be well rinsed with plenty of water.

In the use of this latter re-agent for the purpose indicated, it is important to remember that cyanide of potassium is an energetic poison, which acts not only internally, but externally, by absorption; so that it should never be used when there is the slightest wound or scratch upon the hands.

If this advice be neglected, serious results may follow. It is also advisable that this mixture should only be prepared at the time of using it, and that every vessel which has contained it should be carefully cleaned.

NOTE 6 (*page* 35).

PREPARATION OF THE NITRATE OF SILVER BATH.

Into a graduated glass is placed 1,200 grains of nitrate of silver, upon which is poured 7 ounces of distilled water. When the nitrate is all dissolved, which can be hastened by well stirring with a glass rod, there is to be added 3 drams of an alcoholic solution of iodide of cadmium, containing 10 grains to the ounce; a yellow precipitate is immediately produced. The whole is well agitated together, and left to itself for about one quarter of an hour; more distilled water is then added to make up the quantity of 35 ounces; it is then filtered, and is ready for use.

NOTE 7 (*page* 55).

PREPARATION OF PYROGALLIC ACID.

This substance was discovered by Scheele, who supposed it to be sublimed gallic acid. It contains the elements of gallic acid, minus those of carbonic acid.

Dry and pure pyrogallic acid has the form of lamellar needles, or elongated plates, which are soluble in $2\frac{1}{2}$ parts of water at ordinary temperature, and a little less soluble in ether and alcohol. It has a very bitter taste, and when quite pure, does not redden tincture of litmus. A solution of the pure acid will keep, so to speak, indefinitely, especially if, as is the case when it is prepared for photographic use, an acid be added.

It is of the highest importance to keep pyrogallic acid in stoppered bottles, in the dark; since, unless this be done, it gradually turns brown, through combination with the oxygen of the air, and its properties injured.

Pyrogallic acid is prepared by boiling bruised nut-galls, with seven or eight times their weight of water, for three or four hours, replacing the water as fast as it evaporates. The whole is then thrown on a strainer, and the dirty cake of nut-galls being submitted to powerful pressure, in order to remove all the liquid.

The mixed liquors are then evaporated, first, by rapid ebullition, and afterwards more gently, until they acquire the consistence of an extract, which extract is rendered perfectly dessicated by careful drying in a stove.

This product is then heated for ten or twelve hours in a flat iron vessel, over the top of which is stretched a piece of perforated paper, the whole being covered by a conical paper cap. The vessel is placed on a sand-bath, and the temperature, which should not rise higher than 420° Fahr., is indicated by one, or, still better, two thermometers. And this is the most delicate part of the operation. If the heat be insufficient, no result is obtained, and if it be too much heated, another product is obtained which contains no pyrogallic acid.

Operating in this way, 100 parts of dry extract yield 5 parts of pure pyrogallic acid, and at a more advanced stage of the sublimation, 5 parts of impure acid, which may be purified by resublimation.

Note 8 (*page* 47).

PHOTOGRAPHIC OPTICS.

The reader will readily understand that it is by no means easy in so limited a work as the present to give a clear idea of photographic optics; the observations to be made will therefore be confined to demonstrations of the fundamental principles only of the most advanced of the sciences.

The ultimate constitution or essence of light is entirely unknown. Newton entertained the idea that luminous bodies threw out in all directions exceedingly minute corpuscles, which, on coming into contact with the optic nerve, produced a certain effect, which has been distinguished by the term "luminous effect." It has, however, been since proved that the hypothesis of Newton does not accord with facts, and the view most generally entertained is to the effect, that there exists in space, permeating our atmosphere and all existing bodies, a luminous ether of extreme tenuity, to which luminous bodies have the property of communicating a vibration similar to that which takes place when a spring is suddenly struck; and it is this vibration, communicated with an almost inconceivable rapidity to the optic nerve, which produces the effect which is called light.

It is not certainly known whether this last hypothesis is a correct one, but it is certain that it adapts itself most completely to all the known facts of optics, and that it has materially aided in the discovery of some of them.

It is well known that light travels in straight lines at a rate of 240,000 miles in a second of time.

Reflection of light is that effect which takes place when light falls on a plane mirror.

The light which falls on a mirror from a luminous point is called the *incident ray*, and that which reaches the eye from the mirror is called the *reflected ray*; and if a perpendicular be elevated from the point where these two rays meet, it will be found to make two equal angles, or, in other words, the angle of incidence and the angle of reflection are equal.

Reflection takes place with equal regularity from the surface of curved mirrors, but the image is modified according to the character of the curve, which sometimes enlarges, sometimes diminishes, sometimes reverses the image, and sometimes gives an erect image.

Refraction is that phenomenon which is shown when a stick is plunged into water, by its appearing as though it was broken. It is also a well-known fact, that the bottoms of rivers appear to be a great deal nearer than they really are, and that in consequence great mistakes are sometimes made in judging of their depth; this phenomenon also belongs to refraction. In one word, refraction of light takes place whenever a ray of light deviates from its original course by passing through a transparent body of greater or less density.

In the first place will be considered the most simple case (Fig. 98), in which a ray of light R is incident on a plate of glass A, whose two sides are parallel. At the point where the ray R comes into contact with the glass, imagine a perpendicular line to be drawn, and then observe what happens to the luminous ray. (The perpendicular here mentioned is distinguished by the term *normal.*) Experience has shown that the angle R N, formed by the luminous ray and the normal ray, is greater than the angle formed by the same ray in the interior of the glass with the same normal ray prolonged; but this angle is constant for the same quality or character of glass. On leaving the glass the ray assumes a direction parallel to the direction of the primitive ray.

It follows, then, from what has been stated, that *a ray of light falling on the surface of a piece of glass is refracted both on entering and leaving the glass.*

Figs. 99 and 100 illustrate the same fact, and also show the reflection of a ray from the large face of a triangular prism.

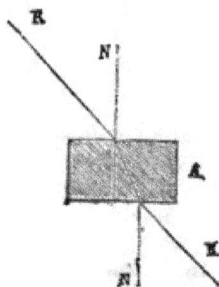

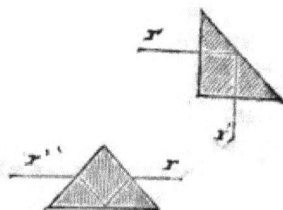

Fig. 99.   Fig. 100.

Fig. 98.

The study of refraction becomes more difficult when the two surfaces of glass, instead of being parallel, are placed at an angle in relation to each other (Fig. 101). Suppose two of these faces A C, A B, and a luminous ray R*o* falling on one of the surfaces. If it were not a prism,

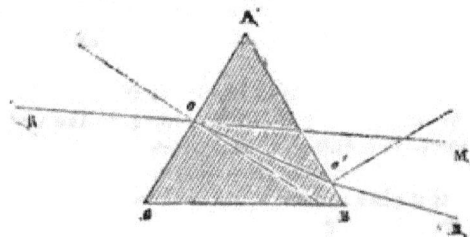

Fig. 101.

the ray would follow the direction R M; but as the ray R*o* is bent, as in the preceding case, it approaches towards the normal in the direction *o o'*, the angle outside the prism being greater than the angle inside. But the reverse happens on the second face A B, at the point *o'*, and the ray leaves in the direction *o' R'* in such a manner that the ray, first impelled in the direction R M, is finally sent in the direction *o' R'*, from which it follows, *that a ray of light is altered in its direction by its passage through a piece of glass, the two sides of which are not parallel.*

Glass lenses are convex discs, ground and polished, having an exterior figure of a spheroidal character. Their principal effect is, that when they are exposed to the sun, the parallel rays of that luminary *r r* (Fig. 102), unite at a certain point, called the *focus*, all the rays converging towards this point. Inversely, a luminous point being placed in the focus *f* of a lens A, emits rays which, on leaving the lens, are parallel

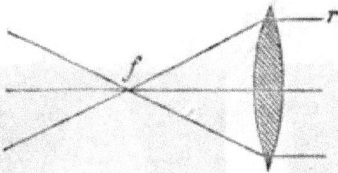

Fig. 102.

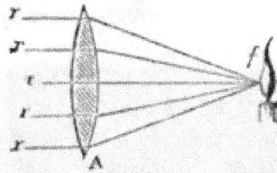

Fig. 103.

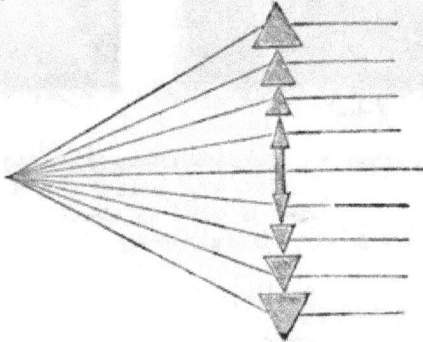

Fig. 104.

r r r r. It therefore follows, that a convex lens is in effect an assemblage of prisms, the inclination of the two faces of the glass from the centre to the circumference being such that the parallel rays undergo an analogous deviation, which causes them to meet at the focus. Fig. 104 is intended to illustrate this proposition.

In *concave* lenses the effect is entirely different. If solar rays *r r*

Fig. 105.

impinge on a concave lens A (Fig. 105), these rays, instead of uniting, disperse themselves in the direction *r' r'*. It is, however, usual to regard the point *f* as the focus, resulting from the ideal prolonging of the rays *r' r'* and *focal length* the distance A *f*.

If refraction consisted only in a simple deviation of the rays of light, it would be a comparatively simple matter, but, unfortunately, it is no so; the ray is not only bent, but decomposed into its primary colours.

This statement may be easily verified by examining a white object on a dark ground through one of the ordinary prisms used for ornamenting chandeliers. The white object will appear fringed with all the colours of the rainbow.

The same phenomenon occurs when a ray of light R (Fig. 101) is allowed to fall on the surface of a prism. The refracted ray o' R' consists not of one white ray, but of seven different colours. Convex lenses produce the same effect. Thus, the solar rays $a\ a$ falling on a convex lens (Fig. 106) do not reunite in one single point, but produce at R a white image, bordered above by red, and below by violet.

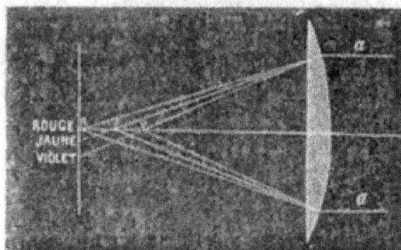

Fig. 106.   Fig. 107.

It is possible, however, to unite two prisms (Fig. 109) in such a way that the rays $b'\ b$ falling thereon, shall leave it at $a'\ a$ void of colour; such a combination of glasses is called an achromatic prism. These prisms are composed of two kinds of glass—crystal, or flint, and crown, or ordinary glass.

There are not only achromatic prisms, but lenses, which yield images free from colour. The curve which a particular kind of glass is to receive in order that the compound lens shall be most effectually achromatised, forms the subject of a mathematical calculation. The curve is then imparted by grinding the glass by hand in a suitable tool.

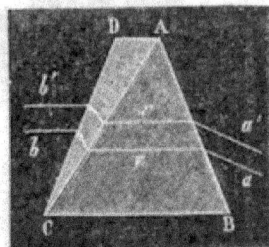

Fig. 108.   Fig. 109.   Fig. 110.   Fig. 111.

There are many different kinds of lenses. Fig. 108 shows three forms of non-achromatic convex lenses, and Fig. 110 these same lenses achromatised. Figs. 109 and 111 concave lenses non-achromatic and achromatised.

In order to explain the application of lenses in photography, it

becomes necessary, in the first place, to describe a very curious phenomenon.

In looking at a well-lighted landscape, it is obvious that every point
of that landscape sends to us a ray of light, since, unless it were so, it
could not be seen at all.

If a hole be bored in the shutter of a perfectly dark room (Fig. 112),
and a sheet of white paper be placed a short distance from the aperture, it
will be seen that, as every point of the view emits luminous rays, a certain

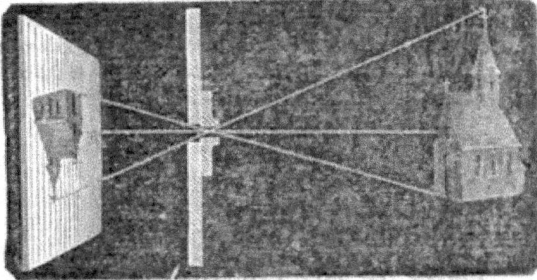

Fig. 112.

number of these rays will pass through the opening into the dark chamber,
and being directed by the sheet of white paper, will produce thereon an
inverted image of the landscape.

If a convex lens be placed in the hole in the shutter, and the sheet of
white paper brought up to the focus of this lens, the image will be increased so much in sharpness that it will seem quite easy to trace the
outline with a pencil, or, indeed, to make a finished drawing. Now, if
the sheet of white paper be replaced by another sheet of paper photo-
graphically prepared with some substances acted on by light—as the
compound of silver, for example—the image of external objects will, in
a longer or shorter time, be depicted thereon.

In Fig. 106 has been shown the remarkable fact, that the two
white rays *a a* become decomposed into rays of various colours; but
what is still more curious is, that if the rays *a a* were red, they would
come to a focus at R; if yellow, nearer the lens at J, and if violet, still
nearer at V. In one word, a lens acts equally only for light of one
colour, and unequally on different colours. Lenses of the same form,
but of different glass, will also act differently on the same light; and it
is precisely on this account that it becomes necessary to make a combination of glasses of such forms that all the colours shall be equally
refracted and reunited in one point.

For this purpose lenses are made and placed very near to each other,
in some instances even united by Canada balsam, one lens, so to speak,
for each colour; generally, however, the combination is confined to two
lenses, uniting only the two principal colours.

In the construction of photographic lenses, the relation between the
material, *i. e.*—the quality and kind of glass—and the form or curve
imparted to it is calculated in such a way as to unite into one focus the
yellow and the violet rays.

Sometimes double and sometimes single objectives are employed.
The double objective is composed of four lenses mounted in brass.

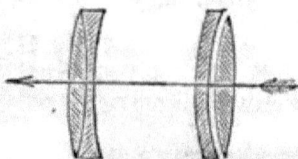

Fig. 53 represents such an one, and
Fig. 113 the arrangement of the lenses.
The point of the arrow is directed towards the object to be taken. This
system of four lenses is arranged in
such a manner as to give a great deal
of light to the image, to the sacrifice, to
some extent, of sharpness. A single
achromatic lens may, however, be employed which, while it gives less light,
wonderfully increases the delicacy of the details.

Fig. 113.

NOTE 9 (*page* 63).

SPOTS ON THE COLLODION FILM.

It is not intended to describe all the kinds of spots which are produced
on the collodion film, but only those which occur most frequently.

Spots are sometimes produced *under* the collodion film, and sometimes
upon the film. The first, always visible before exposure, arise generally
from imperfectly cleaning the plate. In fact, the dust which remains
on the plate are centres of reduction for the iodide of silver constituting
the film, and thus form round spots.

Another source of spots arises from the presence of fatty particles in
the leather used for cleaning, and these produce stains such as shown at
page 63 (fig. 65).

In reference to spots produced *on* the film, they arise very often from
light solid bodies floating in the collodion. It is, therefore, of the highest
importance always to use a collodion which has stood some time.

It happens sometimes that the little crystals of iodo-nitrate which
float in the nitrate bath deposit themselves on the film, and on this
account some photographers pour their silver solution from a bottle, into
which it has been filtered, after sensitising a considerable number of
plates.

At other times the bottom of the plate is riddled with holes. These
are caused by the concentration of the nitrate on account of the plate
having been kept too long between the sensitising and the development.

Sometimes, when the nitrate bath has not been filtered for a long
time, a pellicle of reduced silver is formed, which attaches itself very
firmly to the film. Veins also occur, which are aptly represented by
fig. 66, page 63.—(M. de la Blanchere.)

A similar class of stains sometimes occur, especially on positives, if
the bath contains an excess of alcohol, and they become visible on withdrawing the plate from the sulphate of iron solution.

The stains shown in fig. 67, page 63, arise when too thick a collodion
is used.

Pyrogallic acid often produces spots. If the developer prepared therewith contains too little acetic acid, foggy pictures are the result. If it contains too much, the development proceeds very slowly. But spots occur less frequently in the latter case than the former.

Should too small a quantity of pyrogallic be poured on the plate, stains develop themselves at the corner, and sometimes spread on to the centre; and nothing will remove them.

If the pyrogallic acid developer does not spread itself immediately across the plate, it produces lines which are as irreparable as those stains described in the previous instance. It is also very important to maintain a constant backward and forward motion during development, otherwise a series of little black points of reduced silver will attach themselves to the plate.

Hyposulphite of soda, imperfectly washed away, sometimes agglomerates after the lapse of a certain time, and then produces star-like spots of the form indicated—fig 68, page 63. It is, therefore, of the highest importance to remove the hyposulphite by repeated washing.

It remains to say a few words on fogging, the origin of which is twofold—the first, diffused light, and the second, alkalinity of the bath.

The evil from the first cause arises generally from the inferiority of the yellow glass, which does not completely arrest the passage of the actinic rays; sometimes from the lamp or candle giving off too much white light, from a hole in the camera back, or in the camera itself, &c.

The second of these causes, alkalinity of the bath, is much more rare; it occurs generally in summer time when the weather is very warm. A few drops of acetic acid in the nitrate bath will obviate this defect.

NOTE 10 (*page* 56).

CRYSTALLISABLE ACETIC ACID.

Where crystallisable acetic acid cannot be obtained, its use may, to a certain extent, be dispensed with, by the following plan.

One thousand five hundred grains of caustic potash are dissolved in 35 ounces of distilled water, to which is added 750 grains of powdered litmus, and the blue liquid is decanted into another bottle. This bottle should be kept carefully stoppered. Having obtained some good ordinary acetic acid, called purified pyroligneous acid, and a tube divided into cubic centimetres, two cubic centimetres of the blue solution of potash are poured therein. Now add, drop by drop, some *standard crystallisable acid*, becoming solid when exposed to a temperature of about 40° Fahr.; and after each addition, shake the tube. *A point will be reached at which the blue solution all at once becomes red;* it is at this moment that the operation is completed.

A note is made of the amount of acid which was necessary to change the blue colour of the potass solution; and suppose, for example, this to have been a quarter of a cubic centimetre. Now begin afresh, by mixing the solution of litmus with twice its volume of water, and also the crystallisable acid with a similar proportion. This will allow the observation to be made more accurately as to the quantity of acid required to redden a given quantity of solution of potash. Suppose that it is finally ascertained that 10 cent. cubes of the blue solution require 1 to $1\frac{1}{2}$ cent. cube of the crystallisable acid. Now perform a similar operation with the pyroligneous acid; this being much weaker, it will probably be found that for 10 parts of the blue solution $3\frac{1}{2}$ parts of this acid will be required. The following equation is now made. If $1\frac{1}{2}$ parts of pure acid correspond to $3\frac{1}{2}$ parts of ordinary acid, 10 of pure acid will correspond to x.

From this calculation is deduced the fact, that x is equal to 28; from which it results, that every time 10 grains of crystallisable acid are ordered in a formula, they may be replaced by 28 of ordinary acid: if 5, 14 only will be necessary; if 30, 84 will be wanted, &c. &c.

The hydrometer cannot be used to determine the strength of acetic acid, for its density bears no regular proportion to its saturating power, or, in other words, to the relative quantities of acid and water.

The pure crystallisable acid solidifies between 50° to 60° Fahr. Although the ordinary acid is only a mixture of pure acid and water, and the first solidifies at, say 56° Fahr., and the second at 32°, it does not necessarily follow that the mixture shall solidify above 32°. The pure acid, mixed with three times its volume of water, should, at first sight, solidify at about 40° Fahr., but, in reality, it does not congeal until it is cooled to 36° below the freezing point of water; from which it follows, very weak acid cannot be purified by successive and fractional freezing; and it is only when more concentrated, that this method of purification can be adopted.

INDEX.

James S. Virtue, Printer, City Road, London.

CATALOGUE

OF

Photographic Apparatus and Chemicals,

MANUFACTURED AND SOLD BY

HORNE AND THORNTHWAITE,

121, 122, & 123, NEWGATE STREET,

LONDON, E.C.

LENSES.

	£ s. d.		£ s. d.		£ s. d.
For portraits, 4¼ by 3½	0 16 0	and	1 7 0	and	3 13 6
„ 6½ „ 4¾	2 0 0	„	2 17 6	„	7 0 0
„ 8½ „ 6½	5 0 0	„	8 10 0	„	15 15 0
For landscapes, 6 „ 5	1 15 0	—		—	
„ 7 „ 6	2 12 6	—		—	
„ 9 „ 7	3 13 6	—		—	
„ 12 „ 10	5 5 0	—		—	
„ Stereoscopic	1 13 0	„	2 0 0	„	2 5 0
For Portraits, ditto	3 3 0	„	7 7 0	per pair.	

CAMERAS.

	£ s. d.		£ s. d.
Walnut, expanding . . . from	0 11 6	to	1 10 0
Mahogany, ditto	0 13 6	„	2 12 6
Ditto, best make	0 17 6	„	5 0 0
Ditto, folding, for views . . .	3 0 0	„	17 12 6
Ditto, folding and expanding, ditto .	4 10 0	„	23 0 0
Ditto, Stereoscopic	3 0 0	—	
Ditto, ditto, with lens . . .	6 10 0	—	
Ditto, ditto, Binocular . . .	4 14 6	„	9 9 0
Ditto, ditto, Powell's . . .	7 7 0	„	9 9 0
Ditto, Schaw's Landscape . . .	5 5 0	„	7 7 0

BATHS.

	£ s. d.		£ s. d.
Gutta Percha . . from	0 2 8	to	1 2 6
Ditto, water-tight . .	0 7 6	„	1 16 0
Solid glass	0 6 0	„	1 15 0
Ditto, water-tight . .	1 1 0	„	2 17 6
Porcelain	0 3 6	„	0 14 0
Ditto Well Baths . .	0 3 0	„	0 8 6

GLASS PLATES FOR THE COLLODION PROCESS.

BEST CROWN, EXTRA WHITE SHEET, PURPLE, OPAL,
AND BEST PATENT PLATE.

*For Prices and Sizes see detailed Catalogue, to be had on application, or
by sending Stamp and Address to 122, Newgate Street, London, E.C.*

CRYSTAL MEDIUM.
NIELLO PAPER.
VIGNETTE PLATES.

GROOVED PLATE BOXES FOR GLASS PLATES,
DEAL, MAHOGANY, AND ZINC.

Gutta Percha Trays, Well Baths, and Washing Trays.

PORCELAIN WASHING PANS,
From 5 by 4, 9d. to 24 by 19, £1 1s.

Horne & Thornthwaite's Photographic Colours,
In boxes, 7 colours, 7s. 6d.; 14 colours, 10s. 6d.; 21 colours, 13s.
Separate colours, 6d. each.

CAMERA STANDS.
For Landscapes, 10s. 6d. to £1 11s. 6d.; for the Studio, £1 1s. to £3 13s. 6d.

HEAD RESTS.
In Wood and Metal, 2s., 5s., and 10s. 6d. In Metal, £1 10s. and £3 3s.

STILLS, WITH REFRIGERATORS.
Tin, 1 gal. 21s.; 2 gal. £1 7s. 6d.—Copper, 1 gal. £1 18s.; 2 gal. £2 14s.

STEREOSCOPIC PICTURES,
BRITISH AND FOREIGN, ON GLASS AND PAPER.

STEREOSCOPES,
BREWSTER'S REFRACTING & WHEATSTONE'S REFLECTING.

HORNE & THORNTHWAITE, 121, 122, & 123, NEWGATE STREET.

BACKGROUNDS FOR THE OPERATING ROOM,
£1 10s. to £2 5s.

BRUSHES.	VIEW METERS.
STIRRING RODS.	FOCUSSING EYE PIECES.
FUNNELS.	SPIRIT LAMPS.
SCALES AND WEIGHTS.	CIRCULAR LEVELS.
MEASURES.	GLASS CARBOYS.
FILTERS.	PREPARING BOARDS.
GLASS PESTLES AND MORTARS.	YELLOW GLASS LANTERNS.

GRADUATED COLLODION BOTTLES, frcm 2s. to 3s. 6d.
COLLODION FILTERERS OR POURERS, 6s. 6d. each.

Levelling Stands, 2s. 6d., 4s., 6s.
PRESSURE FRAMES,
6 by 5, 5s. 9d. each.   23 by 21, £1 18s. 6d.

HASKIN'S IMPROVED PRESSURE FRAMES,
From 18s. to £4 10s. per dozen.

MOULDED GLASS DISHES.	PHOTOGRAPHIC PINS.
CARDBOARD FOR CARTES DE VISITE.	AMERICAN CLIPS.
STEREOSCOPIC MOUNTS.	BATH TESTERS.
FOCUSSING CLOTHS.	PLATE HOLDERS.

DR. NORRIS'S DRY COLLODION PLATES.
LEAKE'S PHOTOGRAPHIC TENT.

PHOTOGRAPHIC PAPERS.

Papier Saxe.	Horne & Thornthwaite's waxed iodised.
Papier Rève.	Ditto ditto, sensitive.
Papier Josef.	White filtering.
Whatman's.	Dead black.
Towgood's.	Iodised.
Holingsworth's thin.	Salted.
Balston's.	Albumenised.
Turner's.	
Horne & Thornthwaite's waxed.	

ESTIMATES FOR COMPLETE SETS OF PHOTOGRAPHIC APPARATUS.

HORN FORCEPS.
SILVER PINS.
SILVER WIRE HOOKS.
DROPPING BOTTLES.

WASH BOTTLES.
DEVELOPING GLASSES.
PLATE DRAINERS.
SYPHON WASHING PANS.

ROLLING PRESSES,
From £1 10s. to £12 12s.

ROMALEIAN FRAMES.
PASSE PARTOUTS.
FRAMES FOR PASSE PARTOUTS.
ENGLISH AND AMERICAN MOROCCO
CASES.

MOROCCO FRAMES.
UNION TRAYS.
GILT MATS.
GILT PRESERVERS.
CARDBOARD MOUNTS.

ABSOLUTELY PURE CHEMICALS.

POSITIVE COLLODION.
NEGATIVE DITTO.
ACKLAND'S DITTO.
FOTHERGILL'S DITTO.
CRYSTAL VARNISH.

NEGATIVE VARNISH.
SOEHNÈE DITTO.
CHLOROFORM DITTO.
BLACK (BATES') DITTO.
PLATE-CLEANING SOLUTION.

NITRATE OF SILVER BATH.

BROMO-IODISED COLLODION.

DEVELOPING SOLUTION.

NEW PORTRAIT COLLODION.

White and Green Glass Bottles—Plain and Stoppered.

ILLUSTRATED AND DETAILED **PRICED CATALOGUE**
by post on receipt of one Postage Stamp.

HORNE & THORNTHWAITE, 121, 122, & 123, NEWGATE STREET.

CATALOGUE

OF

RUDIMENTARY, SCIENTIFIC, EDUCATIONAL, AND CLASSICAL WORKS,

FOR COLLEGES, HIGH AND ORDINARY SCHOOLS, AND SELF-INSTRUCTION;

ALSO FOR

MECHANICS' INSTITUTIONS, FREE LIBRARIES, &c. &c.,

PUBLISHED BY

VIRTUE BROTHERS & CO., 1, AMEN CORNER,

PATERNOSTER ROW.

~~~~~~~~~

₃ THE ENTIRE SERIES IS FREELY ILLUSTRATED ON WOOD AND STONE WHERE REQUISITE.

The Public are respectfully informed that the whole of Mr. WEALE'S *Publications, contained in the present Catalogue, have been Purchased by* VIRTUE BROTHERS & Co., *and that all future Orders will be supplied by them at the above address.*

*** *Several additional Volumes, by Popular Authors, are in preparation, and will shortly be ready for delivery.*

RUDIMENTARY SERIES.

2. NATURAL PHILOSOPHY, by Charles Tomlinson .	1s.
3. GEOLOGY, by Major-Gen. Portlock, F.R.S., &c. .	1s. 6d.
6. MECHANICS, by Charles Tomlinson . . .	1s.
12. PNEUMATICS, by Charles Tomlinson . . .	1s.
20, 21. PERSPECTIVE, by George Pyne, 2 vols. in 1 .	2s.
27, 28. PAINTING, The Art of; or, A GRAMMAR OF COLOURING, by George Field, 2 vols. in 1 .	2s.
36, 37, 38, 39. DICTIONARY of the TECHNICAL TERMS used by Architects, Builders, Engineers, Surveyors, &c., 4 vols. in 1 . . . . In cloth boards, 5s.; half morocco, 6s.	4s.
40. GLASS STAINING, by Dr. M. A. Gessert, With an Appendix on the Art of Enamelling . . .	1s.

41. PAINTING ON GLASS, from the German of Emanuel
O. Fromberg . . . . 1s.

69, 70. MUSIC, a Practical Treatise, by C. C. Spencer, Doctor
of Music, 2 vols. in 1 . . . . 2s.

71. THE PIANOFORTE, Instructions for Playing, by C. C.
Spencer, Doctor of Music . . . . . . 1s.

72 to 75*. RECENT FOSSIL SHELLS (A Manual of the
Mollusca), by Samuel P. Woodward, 4 vols. in 1, and
Supplement . . . . 5s. 6d.
In cloth boards, 6s. 6d.; half morocco, 7s. 6d.

83. BOOK-KEEPING, by James Haddon, M.A. . . . 1s.

84. ARITHMETIC, with numerous Examples, by Professor
J. R. Young . . . . 1s. 6d.

84*. KEY TO THE PRECEDING VOLUME, by Professor
J. R. Young . . . . 1s. 6d.

96. ASTRONOMY, POPULAR, by the Rev. Robert Main,
M.R.A.S. . . . . 1s.

101*. WEIGHTS AND MEASURES OF ALL NATIONS;
Weights of Coins, and Divisions of Time; with the
Principles which determine the Rate of Exchange, by
Mr. Woolhouse, F.R.A.S. . . . . . 1s. 6d.

103. INTEGRAL CALCULUS, Examples of, by Prof. J. Hann 1s.

112. DOMESTIC MEDICINE, for the Preservation of Health,
by M. Raspail . . . . 1s. 6d.

131. MILLER'S, FARMER'S, AND MERCHANT'S READY-
RECKONER, showing the Value of any Quantity of
Corn, with the Approximate Value of Mill-stones and
Mill Work . . . . 1s.

(In Preparation.)
PHOTOGRAPHY. A New Manual.

PHYSICAL SCIENCE.

1. CHEMISTRY, by Professor Fownes, F.R.S., including
Agricultural Chemistry, for the use of Farmers . . 1s.

4, 5. MINERALOGY, with a Treatise on Mineral Rocks or
Aggregates, by James Dana, A.M., 2 vols. in 1 . . 2s.

7. ELECTRICITY, an Exposition of the General Principles of
the Science, by Sir William Snow Harris, F.R.S. 1s. 6d.

7*. GALVANISM, ANIMAL AND VOLTAIC ELECTRI-
CITY; A Treatise on the General Principles of Gal-
vanic Science, by Sir William Snow Harris, F.R.S. 1s. 6d.

8, 9, 10. MAGNETISM, Concise Exposition of the General
Principles of Magnetical Science and the Purposes to
which it has been Applied, by the same, 3 vols. in 1 3s. 6d.

BUILDING AND ARCHITECTURE.

VIRTUE BROTHERS & CO., 1, AMEN CORNER.

MACHINERY AND ENGINEERING.

98, 98*. MECHANISM AND THE CONSTRUCTION OF
    MACHINES, by Thomas Baker, C.E.; and TOOLS AND
    MACHINES, by J. Nasmyth, C.E., with 220 Woodcuts 2s. 6d.
114. MACHINERY, its Construction and Working, by C. D.
    Abel, C.E. . . . . 1s. 6d.
115. ILLUSTRATIVE PLATES TO THE ABOVE, 4to. 7s. 6d.
139. THEORY OF THE STEAM ENGINE, by T. Baker, C.E. 1s.

CIVIL ENGINEERING, &c.

13, 14, 15, 15*. CIVIL ENGINEERING, by Henry Law,
    3 vols.; with Supplement by G. R. Burnell, 4 vols. in 1 4s. 6d.
29. DRAINING DISTRICTS AND LANDS, the Art of, by
    G. D. Dempsey, C.E. . . . . . . 1s.
    (With No. 30, DRAINAGE AND SEWAGE OF TOWNS, 2 vols. in 1, 2s. 6d.)
31. WELL-SINKING AND BORING, by John G. Swindell,
    revised by G. R. Burnell, C.E. 1s
46. ROAD-MAKING, the Construction and Repair, by S. C.
    Hughes and H. Law, C.E., and Gen. Sir J. Burgoyne,
    Bart., G.C.B., R.E. . . . . . . 1s. 6d.
60, 61. LAND AND ENGINEERING SURVEYING, by T.
    Baker, C.E., 2 vols. in 1 . . . . . 2s.
63, 64, 65. AGRICULTURAL BUILDINGS, FIELD EN-
    GINES, MACHINERY, and IMPLEMENTS, by G. H.
    Andrews, 3 vols. in 1 . . . . . . 3s.
66. CLAY LANDS AND LOAMY SOILS, by Professor Do-
    naldson, A.E. . . . . 1s.
77*. ECONOMY OF FUEL, by T. S. Prideaux . . . 1s.
80*, 81*. EMBANKING LANDS FROM THE SEA, with
    Examples of actual Embankments and Sea Walls, by
    John Wiggins, F.G.S., 2 vols. in 1 . . . . 2s.
82, 82*. POWER OF WATER, as applied to the Driving of
    Mills, and Giving Motion to Turbines, and other Hy-
    drostatic Machines, by Joseph Glynn, F.R.S., C.E. . 2s.
82**, 83*, 83 bis. COAL GAS, its Manufacture and Distribu-
    tion, by Samuel Hughes, C.E. . . . . 3s.
82***. WATER-WORKS FOR THE SUPPLY OF CITIES
    AND TOWNS, by Samuel Hughes, C.E. . . 3s.
117. SUBTERRANEOUS SURVEYING, & RANGING THE
    LINE without the Magnet, by T. Fenwick, Coal Viewer,
    with Improvements and Additions by T. Baker, C.E. 2s. 6d.
118, 119. CIVIL ENGINEERING IN NORTH AMERICA,
    by D. Stevenson, C.E., 2 vols. in 1 . . . . 3s.
120. HYDRAULIC ENGINEERING, by G. R. Burnell, C.E.,
    2 vols. in 1 . . . . 3s.
140. OUTLINES OF MODERN FARMING, by R. Scott Burn 2s.

121, 122. RIVERS AND TORRENTS, from the Italian of
    Paul Frisi, and a Treatise on NAVIGABLE CANALS,
    AND RIVERS THAT CARRY SAND AND MUD 2s. 6d.
125, 126. COMBUSTION OF COAL, AND THE PREVEN-
    TION OF SMOKE, by Charles Wye Williams, M.I.C.E. 3s.

SHIP-BUILDING AND NAVIGATION.

51, 52, 53. NAVAL ARCHITECTURE, Principles of the
    Science, and their Practical Application to Naval Con-
    struction, by J. Peake, N.A., 3 vols. in 1   .   .   . 3s.
53*. SHIPS AND BOATS FOR OCEAN AND RIVER
    SERVICE, the Principles of Construction, by Captain
    H. A. Sommerfeldt   .   .   .   . 1s.
53**. ATLAS OF 14 PLATES TO THE PRECEDING,
    Drawn to a Scale for Practice   .   .   .   7s. 6d.
54. MASTING, MAST-MAKING, and RIGGING OF SHIPS,
    by R. Kipping, N.A.   .   .   .   .   .   . 1s. 6d.
54*. IRON SHIP-BUILDING, by John Grantham, C.E.   2s. 6d.
54**. ATLAS OF 24 PLATES to the preceding Volume   22s. 6d.
80, 81. MARINE ENGINES AND THE SCREW, by R.
    Murray, C.E., 2 vols. in 1   .   .   .   .   . 2s. 6d.
83 bis. SHIPS AND BOATS, the Principles of Construction,
    by W. Bland, of Hartlip 1s
106. SHIPS' ANCHORS FOR ALL SERVICES, by George
    Cotsell, N.A. 1s. 6d

ARITHMETIC AND MATHEMATICS.

32. MATHEMATICAL INSTRUMENTS, AND THEIR
    USE, by J. F. Heather, M.A.   .   .   .   . 1s.
55, 56. NAVIGATION; the Sailor's Sea Book: How to Keep
    the Log and Work it off, &c.; Law of Storms, and Expla-
    nation of Terms   .   .   .   . 2s.
61*. READY RECKONER for the Measurement of Land, its
    Valuation, and the Price of Labour, by A. Arman,
    Schoolmaster   .   .   .   . 1s. 6d.
76, 77. GEOMETRY, DESCRIPTIVE, with a Theory of Sha-
    dows and Perspective, and a Description of the Principles
    and Practice of Isometrical Projection, by J. F. Heather,
    M.A., 2 vols. in 1   .   .   .   . 2s.
85. EQUATIONAL ARITHMETIC: Questions of Interest,
    Annuities, &c., by W. Hipsley   .   .   .   . 1s.
85*. EQUATIONAL ARITHMETIC: Tables for the Calculation
    of Simple Interest, with Logarithms for Compound Inte-
    rest, and Annuities, by W. Hipsley   .   .   . 1s.

VIRTUE BROTHERS & CO., 1, AMEN CORNER.

86, 87. ALGEBRA, by James Haddon, M.A., 2 vols. in 1 . 2s.
86*, 87*. ELEMENTS OF ALGEBRA, Key to the, by Prof.
Young . . . . 1s. 6d.
88, 89. GEOMETRY, Principles of, by Henry Law, C.E.,
2 vols. in 1 . . . . 2s.
90. GEOMETRY, ANALYTICAL, by James Hann . . 1s.
91, 92. PLANE AND SPHERICAL TRIGONOMETRY, by
Prof. James Hann, 2 vols. in 1 (The two divisions separately, 1s. each) . . . . 2s.
93. MENSURATION, by T. Baker, C.E. . . . . 1s.
94, 95. LOGARITHMS, Tables of; with Tables of Natural
Sines, Co-sines, and Tangents, by H. Law, C.E., 2 vols. in 1 2s. 6d.
97. STATICS AND DYNAMICS, by T. Baker, C.E. . 1s.
99, 100. NAVIGATION AND NAUTICAL ASTRONOMY,
by Professor Young, 2 vols. in 1 . . . . . 2s.
100*. NAVIGATION TABLES, compiled for Practical Use
with the preceding volume . . . . . 1s. 6d.
101. DIFFERENTIAL CALCULUS, by Mr. Woolhouse, F.R.A.S. 1s.
102. INTEGRAL CALCULUS, by H. Cox, M.A. . . . 1s.
104. DIFFERENTIAL CALCULUS, Examples of, by J.
Haddon, M.A. . . . . 1s.
105. ALGEBRA, GEOMETRY, and TRIGONOMETRY, First
Mnemonical Lessons in, by the Rev. T. P. Kirkman, M.A. 1s. 6d.
136. RUDIMENTARY ARITHMETIC, by James Haddon,
M.A., with Additions by A. Arman . . . 1s. 6d.
137. KEY TO THE ABOVE, containing Answers to all the
Questions in that Work, by A. Arman . . 1s. 6d.

MISCELLANEOUS.

50. LAW OF CONTRACTS FOR WORKS AND SERVICES,
by David Gibbons, S.P. . . . . . . 1s.
107. METROPOLITAN BUILDINGS ACT, and THE ME-
TROPOLITAN ACT FOR REGULATING THE
SUPPLY OF GAS, with Notes . . . . 2s. 6d.
108. METROPOLITAN LOCAL MANAGEMENT ACTS 1s. 6d.
108*. METROPOLIS LOCAL MANAGEMENT AMEND-
MENT ACT, 1862: with Notes and Index . . 1s.
110. RECENT LEGISLATIVE ACTS applying to Contractors,
Merchants, and Tradesmen . . . . . 1s.
109. NUISANCES REMOVAL AND DISEASE PREVEN-
TION ACT . . . . 1s.
113. USE OF FIELD ARTILLERY ON SERVICE, by
Lieut.-Col. Hamilton Maxwell, B.A. . . . 1s. 6d.
113*. MEMOIR ON SWORDS, by the same . . . 1s.
83**. CONSTRUCTION OF DOOR LOCKS . . 1s. 6d.

VIRTUE BROTHERS & CO., 1, AMEN CORNER.

NEW SERIES OF EDUCATIONAL WORKS.

[This Series is kept in three styles of binding—the prices of each are given in columns at the end of the lines.]

HISTORIES, GRAMMARS, AND DICTIONARIES.	Limp.	Cloth Boards.	Half Morocco.
	s. d.	s. d.	s. d.
1, 2, 3, 4. CONSTITUTIONAL HISTORY OF England, by W. D. Hamilton . . .	4 0	5 0	5 6
5, 6. OUTLINES OF THE HISTORY OF Greece, by E. Levien, M.A., 2 vols. in 1 .	2 6	3 6	4 0
7, 8. OUTLINES OF THE HISTORY OF Rome, by the same, 2 vols. in 1 . . .	2 6	3 6	4 0
9, 10. CHRONOLOGY OF CIVIL AND Ecclesiastical History, Literature, Art, and Civilization, from the earliest period to the present, 2 vols. in 1	2 6	3 6	4 0
11. GRAMMAR OF THE ENGLISH LANGUAGE, by Hyde Clarke, D.C.L. . .	1 0		
11*. HAND-BOOK OF COMPARATIVE Philology, by the same	1 0		
12, 13. DICTIONARY OF THE ENGLISH Language.—A new Dictionary of the English Tongue, as spoken and written; above 100,000 words, or 50,000 more than in any existing work, by the same, 3 vols. in 1 .	3 6	4 6	5 0
——————, with the Grammar .		5 6	6 0
14. GRAMMAR OF THE GREEK LANGUAGE, by H. C. Hamilton . . .	1 0		
15, 16. DICTIONARY OF THE GREEK AND English Languages, by H. R. Hamilton, 2 vols. in 1	2 0		
17, 18. DICTIONARY OF THE ENGLISH and Greek Languages, by the same, 2 vols. in 1	2 0		
—————— GREEK AND ENGLISH and English and Greek, 4 vols. in 1 . .		5 0	5 6
——————, with the Greek Grammar .		6 0	6 6
19. GRAMMAR OF THE LATIN LANGUAGE, by the Rev. T. Goodwin, A.B. . . .	1 0		
20, 21. DICTIONARY OF THE LATIN AND English Languages, by the same. Vol. I.	2 0		
22, 23. DICTIONARY OF THE ENGLISH and Latin Languages, by the same. Vol. II.	1 6		
——————, 2 vols. in 1		4 6	5 0
——————, with the Latin Grammar .		5 6	6 0
24. GRAMMAR OF THE FRENCH LANGUAGE, by the Lecturer at Besançon . .	1 0		

VIRTUE BROTHERS & CO., 1, AMEN CORNER.

HISTORIES, GRAMMARS, AND DICTIONARIES.	Limp.	Cloth Boards.	Half Morocco.
	s. d.	s. d.	s. d.
25. DICTIONARY OF THE FRENCH AND English Languages, by A. Elwes. Vol. I.	1 0		
26. DICTIONARY OF THE ENGLISH AND French Languages, by the same. Vol. II.	1 6		
———, 2 vols. in 1		3 6	4 0
———, with the French Grammar .		4 6	5 0
27. GRAMMAR OF THE ITALIAN LANGUAGE, by the same	1 0		
28, 29. DICTIONARY OF THE ITALIAN, English, and French Languages, by the same. Vol. I.	2 0		
30, 31. DICTIONARY OF THE ENGLISH, Italian, and French Languages, by the same. Vol. II.	2 0		
32, 33. DICTIONARY OF THE FRENCH, Italian, and English Languages, by the same. Vol. III.	2 0		
———, 3 vols. in 1 . . .		7 6	8 6
———, with the Italian Grammar .		8 6	9 6
34. GRAMMAR OF THE SPANISH LANGUAGE, by the same . . .	1 0		
35, 36, 37, 38. DICTIONARY OF THE Spanish and English Languages, by the same, 4 vols. in 1	4 0	5 0	5 6
———, with the Spanish Grammar .		6 0	6 6
39. GRAMMAR OF THE GERMAN LANGUAGE, by the Lecturer at Besançon . .	1 0		
40. CLASSICAL GERMAN READER, from the best authors, by the same . .	1 0		
41, 42, 43. DICTIONARIES OF THE ENGLISH, German, and French Languages, by N. E. Hamilton, 3 vols., separately 1s. each .	3 0	4 0	4 6
———, with the German Grammar .		5 0	5 6
44, 45. DICTIONARY OF THE HEBREW and English Languages, containing the Biblical and Rabbinical words, 2 vols. (together with the Grammar, which may be had separately for 1s.) by Dr. BRESSLAU, Hebrew Professor	7 0		
46. DICTIONARY OF THE ENGLISH AND Hebrew Languages. Vol. III. to complete, by the same	3 0		
———, 3 vols. as 2 . . .		12 0	14 0
47. FRENCH AND ENGLISH PHRASE Book	1 0	1 6	

Now in the course of Publication.

GREEK AND LATIN CLASSICS.

A Series of Volumes containing the principal Greek and Latin Authors, accompanied by Explanatory Notes in English, principally selected from the best and most recent German Commentators, and comprising all those Works that are essential for the Scholar and the Pupil, and applicable for the Universities of Oxford, Cambridge, Edinburgh, Glasgow, Aberdeen, and Dublin; the Colleges at Belfast, Cork, Galway, Winchester, and Eton; and the great Schools at Harrow, Rugby, &c.—also for Private Tuition and Instruction, and for the Library.

LATIN SERIES.

1. A New LATIN DELECTUS, Extracts from Classical Authors, with Vocabularies and Explanatory Notes . 1s.
2. CÆSAR'S COMMENTARIES on the GALLIC WAR; with Grammatical and Explanatory Notes in English, and a Geographical Index . . . . . 2s.
3. CORNELIUS NEPOS; with English Notes, &c. . . 1s.
4. VIRGIL. The Georgics, Bucolics, and doubtful Works; with English Notes . . . . 1s.
5. VIRGIL'S ÆNEID (on the same plan as the preceding) . 2s.
6. HORACE. Odes and Epodes; with English Notes, and Analysis and Explanation of the Metres . . . 1s.
7. HORACE. Satires and Epistles; with English Notes, &c. 1s. 6d.
8. SALLUST. Conspiracy of Catiline, Jugurthine War . 1s. 6d.
9. TERENCE. Andrea and Heautontimorumenos . 1s. 6d.
10. TERENCE. Phormio, Adelphi, and Hecyra . . . 2s.
14. CICERO. De Amicitia, de Senectute, and Brutus . . 2s.
16. LIVY. Books i. to v. in two parts . . . . . 3s.
17. LIVY. Books xxi. and xxii. . . . . . . 1s.
19. Selections from TIBULLUS, OVID, and PROPERTIUS . 2s.
20. Selections from SUETONIUS and the later Latin Writers . 2s.

Preparing for Press.

11. CICERO. Orations against Catiline, for Sulla, for Archias, and for the Manilian Law.
12. CICERO. First and Second Philippics; Orations for Milo, for Marcellus, &c.
13. CICERO. De Officiis.
15. JUVENAL and PERSIUS. (The indelicate passages expunged.)
18. TACITUS. Agricola; Germania; and Annals, Book i.

VIRTUE BROTHERS & CO., 1, AMEN CORNER.

GREEK SERIES,

ON A SIMILAR PLAN TO THE LATIN SERIES.

1. INTRODUCTORY GREEK READER. On the same
plan as the Latin Reader . . . . . . 1s.
2. XENOPHON. Anabasis, i. ii. iii. . . . . 1s.
3. XENOPHON. Anabasis, iv. v. vi. vii. . . . 1s.
4. LUCIAN. Select Dialogues . . . . . 1s.
5. HOMER. Iliad, i. to vi. . . . . . 1s. 6d.
6. HOMER. Iliad, vii. to xii. . . . . . 1s. 6d.
7. HOMER. Iliad, xiii. to xviii. . . . . 1s. 6d.
8. HOMER. Iliad, xix. to xxiv. . . . . 1s. 6d.
9. HOMER. Odyssey, i. to vi. . . . . 1s. 6d.
10. HOMER. Odyssey, vii. to xii. . . . 1s. 6d.
11. HOMER. Odyssey, xiii. to xviii. . . . 1s. 6d.
12. HOMER. Odyssey, xix. to xxiv.; and Hymns . . 2s.
13. PLATO. Apology, Crito, and Phædo . . . 2s.
14. HERODOTUS, i. ii. . . . . . . 1s. 6d.
15. HERODOTUS, iii. iv. . . . . . 1s. 6d.
16. HERODOTUS, v. vi. and part of vii. . . . 1s. 6d.
17. HERODOTUS. Remainder of vii. viii. and ix. . . 1s. 6d.
18. SOPHOCLES; Œdipus Rex . . . . . 1s.
20. SOPHOCLES; Antigone . . . . . 2s.
23, 24. EURIPIDES; Hecuba and Medea . . . 1s. 6d.
26. EURIPIDES; Alcestis . . . . . 1s.
30. ÆSCHYLUS; Prometheus Vinctus . . . . 1s.
41. THUCYDIDES, i. . . . . . . 1s.

Preparing for Press.

19. SOPHOCLES; Œdipus Colonæus.
21. SOPHOCLES; Ajax.
22. SOPHOCLES; Philoctetes.
25. EURIPIDES; Hippolytus.
27. EURIPIDES; Orestes.
28. EURIPIDES. Extracts from the remaining plays.
29. SOPHOCLES. Extracts from the remaining plays.
31. ÆSCHYLUS; Persæ.
32. ÆSCHYLUS; Septem contra Thebes.
33. ÆSCHYLUS; Choëphoræ.
34. ÆSCHYLUS; Eumenides.
35. ÆSCHYLUS; Agamemnon.
36. ÆSCHYLUS; Supplices.
37. PLUTARCH; Select Lives.
38. ARISTOPHANES; Clouds.
39. ARISTOPHANES; Frogs.
40. ARISTOPHANES; Selections from the remaining Comedies.
42. THUCYDIDES, ii.
43. THEOCRITUS; Select Idyls.
44. PINDAR.
45. ISOCRATES.
46. HESIOD.

VIRTUE BROTHERS & CO., 1, AMEN CORNER.

LE PAGE'S FRENCH COURSE.

"The sale of many thousands, and the almost universal adoption of these clever little books by M. LE PAGE, sufficiently prove the public approbation of his plan of teaching French, which is in accordance with the natural operation of a child learning its native language."

LE PAGE'S FRENCH SCHOOL. Part I. L'ECHO DE PARIS; being a Selection of Familiar Phrases which a person would hear daily if living in France. Price 3s. 6d. cloth.

LE PAGE'S FRENCH SCHOOL. Part II. THE GIFT OF FLUENCY IN FRENCH CONVERSATION. With Notes. Price 2s. 6d. cloth.

LE PAGE'S FRENCH SCHOOL. PART III. THE LAST STEP TO FRENCH; with the Versification. Price 2s. 6d. cloth.

LE PAGE'S FRENCH MASTER FOR BEGINNERS; or, EASY LESSONS IN FRENCH. Price 2s. 6d. cloth.

LE PAGE'S PETIT CAUSEUR; or, FIRST CHATTERINGS IN FRENCH. Being a Key to the Gift of French Conversation. Price 1s. 6d.

LE PAGE'S NICETIES OF PARISIAN PRONUNCIATION. Price 6d.

LE PAGE'S JUVENILE TREASURY OF FRENCH CONVERSATION. With the English before the French. Price 3s. cloth.

LE PAGE'S KEY TO L'ECHO DE PARIS. Price 1s.

LE PAGE'S FRENCH PROMPTER. A HANDBOOK FOR TRAVELLERS on the Continent and Students at Home. Price 4s. cloth.

LE PAGE'S READY GUIDE TO FRENCH COMPOSITION. FRENCH Grammar by Examples, giving Models as Leading-strings throughout Accidence and Syntax. Price 3s. 6d. cloth.

Fifth Edition, improved and corrected, in 1 vol. 12mo., neatly bound, price 2s. 6d.,

TATE'S ELEMENTS OF COMMERCIAL ARITHMETIC;

Containing a Minute Investigation of the Principles of the Science, and their general application to Commercial Calculations, in accordance with the present Monetary System of the World. By WILLIAM TATE, Principal of the City of London Establishment for finishing Young Men for Mercantile and Banking Pursuits.

KEY TO THE ABOVE, 12mo., bound, 3s. 6d.

VIRTUE BROTHERS & CO., 1, AMEN CORNER.

www.ingramcontent.com/pod-product-compliance
Lightning Source LLC
Chambersburg PA
CBHW021811190326
41518CB00007B/543